Applied Modeling Techniques and Data Analysis 2

Big Data, Artificial Intelligence and Data Analysis Set
coordinated by
Jacques Janssen

Volume 8

Applied Modeling Techniques and Data Analysis 2

Financial, Demographic, Stochastic and Statistical Models and Methods

Edited by

Yannis Dimotikalis
Alex Karagrigoriou
Christina Parpoula
Christos H. Skiadas

ISTE

WILEY

First published 2021 in Great Britain and the United States by ISTE Ltd and John Wiley & Sons, Inc.

ISTE Ltd
27-37 St George's Road
London SW19 4EU
UK

www.iste.co.uk

John Wiley & Sons, Inc.
111 River Street
Hoboken, NJ 07030
USA

www.wiley.com

Library of Congress Control Number: 2020951002

British Library Cataloguing-in-Publication Data
A CIP record for this book is available from the British Library
ISBN 978-1-78630-674-6

Contents

Preface

Data analysis as an area of importance has grown exponentially, especially during the past couple of decades. This can be attributed to a rapidly growing technology industry and the wide applicability of computational techniques, in conjunction with new advances in analytic tools. Modeling enables analysts to apply various statistical models to the data they are investigating, to identify relationships between variables, to make predictions about future sets of data, as well as to understand, interpret and visualize the extracted information more strategically. Many new research results have recently been developed and published and many more are developing and in progress at the present time. The topic is also widely presented at many international scientific conferences and workshops. This being the case, the need for the literature that addresses this is self-evident. This book includes the most recent advances on the topic. As a result, on one hand, it unifies in a single volume all new theoretical and methodological issues and, on the other, introduces new directions in the field of applied data analysis and modeling, which are expected to further grow the applicability of data analysis methods and modeling techniques.

This book is a collective work by a number of leading scientists, analysts, engineers, mathematicians and statisticians, who have been working on the front end of data analysis. The chapters included in this collective volume represent a cross-section of current concerns and research interests in the above-mentioned scientific areas. This volume is divided into two parts with a total of 17 chapters in a form that provides the reader with both theoretical and applied information on data analysis methods, models and techniques, along with appropriate applications.

Part 1 focuses on financial and demographic modeling techniques and includes nine chapters: Chapter 1, “Data Mining Application Issues in the Taxpayer Selection

Process", by *Mauro Barone, Stefano Pisani and Andrea Spingola*; Chapter 2, "Asymptotics of Implied Volatility in the Gatheral Double Stochastic Volatility Model", by *Mohammed Albuhayri, Anatoliy Malyarenko, Sergei Silvestrov, Ying Ni, Christopher Engström, Finnan Tewolde and Jiahui Zhang*; Chapter 3, "New Dividend Strategies", by *Ekaterina Bulinskaya*; Chapter 4, "Introduction of Reserves in Self-adjusting Steering the Parameters of a Pay-As-You-Go Pension Plan", by *Keivan Diakite, Abderrahim Oulidi and Pierre Devolder*; Chapter 5, "Forecasting Stochastic Volatility for Exchange Rates using EWMA", by *Jean-Paul Murara, Anatoliy Malyarenko, Milica Rancic and Sergei Silvestrov*; Chapter 6, "An Arbitrage-free Large Market Model for Forward Spread Curves", by *Hossein Nohrouzian, Ying Ni and Anatoliy Malyarenko*; Chapter 7, "Estimating the Healthy Life Expectancy (HLE) in the Far Past: The Case of Sweden (1751–2016) with Forecasts to 2060", by *Christos H. Skiadas and Charilaos Skiadas*; Chapter 8, "Vaccination Coverage Against Seasonal Influenza of Workers in the Primary Health Care Units in the Prefecture of Chania", by *Aggeliki Maragkaki and George Matalliotakis*; Chapter 9, "Some Remarks on the Coronavirus Pandemic in Europe", by *Konstantinos N. Zafeiris and Marianna Koukli.*

Part 2 covers the area of applied stochastic and statistical models and methods and comprises eight chapters: Chapter 10, "The Double Flexible Dirichlet: A Structured Mixture Model for Compositional Data", by *Roberto Ascari, Sonia Migliorati and Andrea Ongaro*; Chapter 11, "Quantization of Transformed Lévy Measures", by *Mark Anthony Caruana*; Chapter 12, "A Flexible Mixture Regression Model for Bounded Multivariate Responses", by *Agnese M. Di Brisco and Sonia Migliorati*; Chapter 13, "On Asymptotic Structure of the Critical Galton–Watson Branching Processes with Infinite Variance and Allowing Immigration", by *Azam A. Imomov and Erkin E. Tukhtaev*; Chapter 14, "Properties of the Extreme Points of the Joint Eigenvalue Probability Density Function of the Wishart Matrix", by *Asaph Keikara Muhumuza, Karl Lundengård, Sergei Silvestrov, John Magero Mango and Godwin Kakuba*; Chapter 15, "Forecast Uncertainty of the Weighted TAR Predictor", by *Francesco Giordano and Marcella Niglio*; Chapter 16, "Revisiting Transitions Between Superstatistics", by *Petr Jizba and Martin Prokš*; Chapter 17, "Research on Retrial Queue with Two-Way Communication in a Diffusion Environment", by *Viacheslav Vavilov*.

We wish to thank all the authors for their insights and excellent contributions to this book. We would like to acknowledge the assistance of all those involved in the reviewing process of this book, without whose support this could not have been successfully completed. Finally, we wish to express our thanks to the secretariat and, of course, the publishers. It was a great pleasure to work with them in bringing to life this collective volume.

Yannis DIMOTIKALIS
Crete, Greece
Alex KARAGRIGORIOU
Samos, Greece
Christina PARPOULA
Athens, Greece
Christos H. SKIADAS
Athens, Greece

December 2020

PART 1

Financial and Demographic Modeling Techniques

1

Data Mining Application Issues in the Taxpayer Selection Process

This chapter provides a data analysis framework designed to build an effective learning scheme aimed at improving the Italian Revenue Agency's ability to identify non-compliant taxpayers, with special regard to self-employed individuals allowed to keep simplified registers. Our procedure involves building two C4.5 decision trees, both trained and validated on a sample of 8,000 audited taxpayers, but predicting two different class values, based on two different predictive attribute sets. That is, the first model is built in order to identify the most likely non-compliant taxpayers, while the second identifies the ones that are are less likely to pay the additional due tax bill. This twofold selection process target is needed in order to maximize the overall audit effectiveness. Once both models are in place, the taxpayer selection process will be held in such a way that businesses will only be audited if they are judged as worthy by both models. This methodology will soon be validated on real cases: that is, a sample of taxpayers will be selected according to the classification criteria developed in this chapter and will subsequently be involved in some audit processes.

1.1. Introduction

Fraud detection systems are designed to automate and help reduce the manual parts of a screening/checking process (Phua *et al.* 2005). Data mining plays an important role in fraud detection as it is often applied to extract fraudulent behavior profiles hidden behind large quantities of data and, thus, may be useful in decision support systems for planning effective audit strategies. Indeed, huge amounts of resources (to put it bluntly, money) may be recovered from well-targeted audits. This explains the increasing interest and investments of both governments and fiscal agencies

---

Chapter written by Mauro BARONE, Stefano PISANI and Andrea SPINGOLA.

in intelligent systems for audit planning. The Italian Revenue Agency (hereafter, IRA) itself has been studying data mining application techniques in order to detect tax evasion, focusing, for instance, on the tax credit system, supposed to support investments in disadvantaged areas (de Sisti and Pisani 2007), on fraud related to credit mechanisms, with regard to value-added tax – a tax that is levied on the price of a product or service at each stage of production, distribution or sale to the end consumer, except where a business is the end consumer, which will reclaim this input value (Basta *et al.* 2009) and on income indicators audits (Barone *et al.* 2017).

This chapter contributes to the empirical literature on the development of classification models applied to the tax evasion field, presenting a case study that focuses on a dataset of 8,000 audited taxpayers on the fiscal year 2012, each of them described by a set of features, concerning, among others, their tax returns, their properties and their tax notice. [1]

In this context, all the taxpayers are in some way "unfaithful", since all of them have received a tax notice that somehow rectified the tax return they had filed. Thus, the predictive analysis tool we develop is designed to find patterns in data that may help tax offices recognize only the riskiest taxpayers' profiles.

Evidence on data at hand shows that our first model, which is described in detail later, is able to distinguish the taxpayers who are worthy of closer investigation from those who are not. [2]

However, by defining the class value as a function of the higher due taxes, we satisfy the need of focusing on the taxpayers who are more likely to be "significant" tax evaders, but we do not ensure an efficient collection of their tax debt. Indeed, data shows that as the tax bill increases, the number of coercive collection procedures put in place also increases. Unfortunately, these procedures are highly inefficient, as they are able to only collect about 5% of the overall credits claimed against the audited taxpayers (Italian Court of Auditors 2016). As a result, the tax authorities' ability to collect the due taxes may be jeopardized.

Further analysis is thus devoted to finding a way to discover, among the "significant" evaders, the most solvent ones. We recall that the 2018–2020 Agreement between the IRA and the Ministry of Finance states that audit effectiveness is measured, among others, by an indicator that is simply equal to the sum of the collected due taxes which *summarizes the effectiveness of the IRA's efforts to tackle tax evasion* (Ministry of Economy and Finance – IRA Agreement for 2018–2010 2018). This is a reasonable indicator because the ordinary activities taken in the fight

1 A tax notice is a formal written act through which tax authorities assess a higher due taxable income with respect to the declared one.

2 Data analyses are performed using WEKA – the data mining workbench developed at Waikato University in Hamilton, New Zealand, released under the GNU GPL license.

against tax evasion are crucial from the State budget point of view, because public expenditures (i.e. public services) strictly depend on the amount of public revenue. Of course, fraud and other incorrect fiscal behaviors may be tackled, even though no tax collection is guaranteed, in order to reach the maximum tax compliance. Such extra activities may also be jointly conducted with the Finance Guard or the Public Prosecutor if tax offenses arise.

Therefore, to tackle our second problem, i.e. to guarantee a certain degree of due tax collection, a trivial fact that we start from is that a taxpayer with no properties will not be willing to pay his dues, whereas if he had something to lose (a home or a car that could be seized), then, if the IRA's claim is right, it is more probable that he might reach an agreement with the tax authorities.

Therefore, a second model only focusing on a few features indicating whether the taxpayer owned some kind of assets or not is built, in order to predict each tax notice's final status (in this case, we only distinguish between statuses ending with an enforced recovery proceeding and statuses where such enforced recovery proceedings do not take place). Once both models are available, the taxpayer selection process is held in such a way that businesses will only be audited if they are judged as worthy by both models.

The key feature of our procedure is the twofold selection process target, needed to maximize the IRA's audit processes' effectiveness. The methodology we suggest will soon be validated in real cases i.e. a sample of taxpayers will be selected according to the classification criteria developed in this chapter and will be subsequently involved in some audit processes.

1.2. Materials and methods

1.2.1. *Data*

Data on hand refers to a sample of 8,028 audited self-employed individuals for fiscal year 2012, each described by a set of features, concerning, among others, their tax returns, their properties and their tax notice.[3]

Just for descriptive purposes, we can depict the statistical distribution of the revenues achieved by the businesses in our sample, grouped in classes (in thousands of euros), in Figure 1.1.

Most of our dataset is made up of small-sized taxpayers, of which almost 50% show revenues lower than € 75,000 per year and only 4% higher than € 500,000, with a sample average of € 146,348.

3 The IRA sent a total of 59,269 tax notices concerning fiscal year 2012 to self-employed individuals allowed to keep simplified registers, so we can manage a quite significant sample.

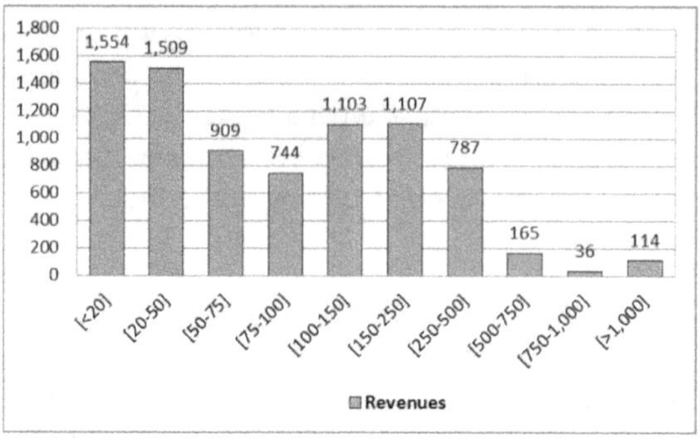

Figure 1.1. *Revenues distribution*

For each taxpayer in the dataset, both his tax notice status and the additional due taxes (i.e. the additional requested tax amount) are known.

Here comes the first problem that needs to be tackled: the additional due tax is a numeric attribute which measures the seriousness of the taxpayer's tax evasion, whereas our algorithms, as we will show later on, need categorical values in order to predict. Thus, we cannot directly use the additional due taxes, but we need to define a *class variable* and decide both which values it will take and how to map each numeric value referred to the additional due taxes into such categorical values.

1.2.2. *Interesting taxpayers*

We must define a function $f(x)$ which associates, to each element x in the dataset, a categorical value that shows its fraud risk degree and represents the *class* our first model will try to predict. Of course, a function that labels all the taxpayers in the dataset as *tax evaders* would be useless. Thus, a distinction needs to be drawn between serious tax evasion cases and those that are less relevant. To this purpose, we somehow follow (Basta *et al.* 2009) and choose to divide the taxpayers into two groups, the *interesting* ones and the *not interesting* ones, from the tax administration point of view (to a certain extent, *interesting* stands for "*it might be interesting for the tax administration to go and check what's going on* ..."), based on two criteria: profitability (i.e. the ability to identify the most serious cases of tax evasion, independently from all other factors) and fairness (i.e. the ability to identify the most serious cases of tax evasion, with respect to the taxpayer's turnover).

Honest taxpayers are treated as *not interesting* taxpayers, even though this label is used to indicate moderate tax evasion cases. We are somehow forced to use this approximation since we only have data on taxpayers who received a tax notice, and not on taxpayers for which an audit process may have been closed without qualifications, or may have not even been started.

Therefore, in order to take the *profitability issue* into account, we define a new variable, called the *tax claim*, which represents the higher assessed taxes if the tax notice stage is still *open*, or the higher settled taxes if the stage status is *definitive*. Note that the higher assessed tax could be different from the higher settled tax, because the IRA and the taxpayer, while reaching an agreement, can both reconsider their positions. The *tax claim* distribution grouped in classes (again, in thousands of euros) is shown in Figure 1.2.

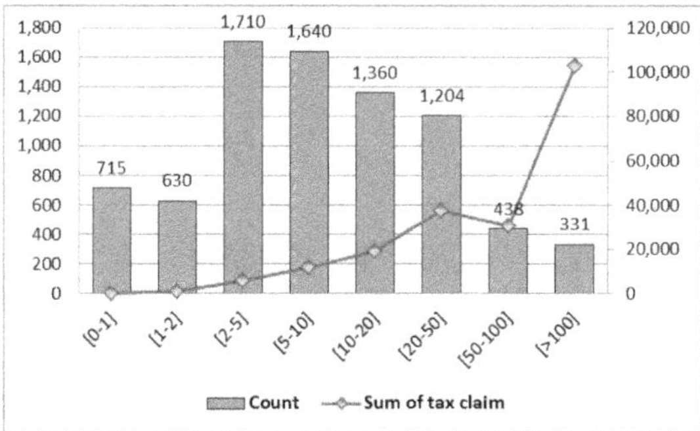

Figure 1.2. *Tax claim distribution. For a color version of this figure, see www.iste.co.uk/dimotikalis/analysis2.zip*

The left vertical axis is related to the *tax claim* distribution, grouped in the classes shown on the horizontal axis; the right vertical axis, on the contrary, sums up the monetary *tax claim* amount that arises from each group (in thousands of euro). Therefore, as it can easily be seen, the 331 most profitable tax notices (12% of the total) account for almost half of the tax revenue arising from our dataset.

The *fairness* criterion is then introduced to address the audit process, even towards smaller firms (which usually are charged smaller amounts of due income taxes), and it is useful as it allows the tax authorities to not discriminate against taxpayers on the basis of their turnover and introduces a deterrent effect which improves the overall *tax compliance*.

Therefore, we define another variable, called Z, which takes into account, for each taxpayer, both his turnover and revenues, and compares them to the *tax claim*. More formally, both of the ratios $\frac{tax\ claim}{turnover}$ and $\frac{tax\ claim}{revenues}$ are computed. Then, the minimum between these two ratios and 1 is taken. That is, the variable Z value, which thus ranges from 0 to 1.

Now, for both *tax claim* (TC) and Z, we calculate the 25th percentile (Q_1), the median value (Q_2) and the 75th percentile (Q_3). We then state that a taxpayer may be considered *interesting* if he satisfies one of the following conditions:

$$Q_1 \leq TC < Q_2 \quad and \quad Z \geq Q_3$$

$$TC \geq Q_2 \quad and \quad Z \geq Q_2$$

$$TC \geq Q_3 \quad and \quad Z < Q_2$$

The three above-mentioned rules can be represented as in Figure 1.3.

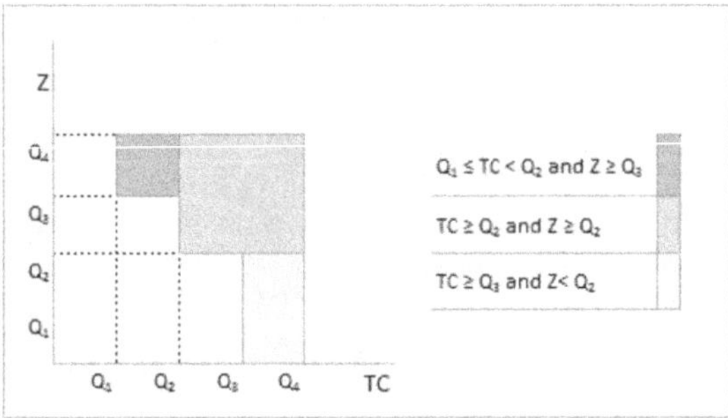

Figure 1.3. *Determining interesting and not interesting taxpayers. For a color version of this figure, see www.iste.co.uk/dimotikalis/analysis2.zip*

Once the population of our dataset is entirely divided into *interesting* and *not interesting* taxpayers, we can see from Table 1.1 that the *interesting* ones are far more *profitable* than the others (*tax claim* values are in thousands of euros). A machine learning tool able to distinguish these two kinds of taxpayers fairly well would then be very useful.

Our first model task will then be that of identifying, with a certain confidence degree, the taxpayers who are more likely to have evaded (both in absolute terms and as a percentage of revenues or turnover).

The literature on tax fraud detection, although using different methods and algorithms, is usually only concerned about this issue, i.e. in finding the best way to identify the most relevant cases of tax evasion (Bonchi *et al.* 1999; Wu *et al.* 2012; Gonzalez and J.D. Velasquez 2013; de Roux *et al.* 2018).

There is another crucial issue that has to be taken into account, i.e. the effective tax authorities' ability to collect the tax debt arising from the tax notices sent to all of the unfaithful taxpayers.

	Not interesting			Interesting		
Tax claim	Num	Total tax claim	Average	Num	Total tax claim	Average
[0 - 1]	736	322	0.44	0	0	0.00
[1 - 2]	631	942	1.49	0	0	0.00
[2 - 5]	1,607	5,409	3.37	138	563	4.08
[5 - 10]	1,127	7,727	6.86	517	4,157	8.04
[10 - 20]	446	5,911	13.25	902	13,139	14.57
[20 - 50]	0	0	0.00	1,164	36,056	30.98
[50 - 100]	0	0	0.00	433	30,055	69.41
[100+]	0	0	0.00	327	101,987	311.89
Total	4,547	20,311	4.47	3,481	185,957	53.42

Table 1.1. *Tax claim, interesting and not interesting taxpayers*

1.2.3. *Enforced tax recovery proceedings*

What happens if a taxpayer does not spontaneously pay the additional tax amount he is charged? Well, after a while, coercive collection procedures will be deployed by the tax authorities. However, as we have seen above, these procedures are highly ineffective, as they only collect about the 5% of the overall credits claimed against the audited taxpayers.

Indeed, data shows that coercive procedures take place in almost 40% of cases, although its distribution is not uniform: they are more frequent if the tax bill is high, as reported in Table 1.2 (again, *tax claim* values are in thousands of euros).

Tax claim	Coercive procedures		Total
	No	Yes	
[0 - 1]	578	158	736
[1 - 2]	476	155	631
[2 - 5]	1,268	477	1,745
[5 - 10]	1,072	572	1,644
[10 - 20]	745	603	1,348
[20 - 50]	511	653	1,164
[50 - 100]	159	274	433
[100+]	90	237	327
Total	4,899	3,129	8,028

Table 1.2. *Number of coercive procedures per tax claim interval*

Table 1.2 is actually a double frequency table, which can be used to investigate the existing relationship between the two categorical variables, *Coercive procedures* and *Tax claim* (they both take on values that are labels). Recall that given characters X and Y, X is independent from Y if for all Y values, the relative distribution of X does not change. Therefore, a quick glance at Table 1.2 shows that *Coercive procedures* depend on the values taken by *Tax claim*.

In a more formal way, following the Openstax (2013) notation, we could also perform a test of independence for these variables, by using the well-known test statistic for a test of independence:

$$\chi^2 = \sum_{i.j} \frac{(O - E)^2}{E}$$

where O is the observed value, E is the expected value, calculated as (row total)(column total) over total number surveyed.

Given the values in Table 1.2, the test would let us reject the hypothesis of the two variables being independent at a 1% level of significance: therefore, from the data, there is sufficient evidence to conclude that *Coercive procedures* are dependent on the *Tax claim* level.

It is easy to calculate, from Table 1.2, for each *tax claim* interval, the total coercive procedures rate, the tax notices rate and the coercive procedures within that *tax claim* interval rate (all of these ratios are depicted in Figure 1.4).

A close look at Figure 1.4 shows that until the tax claim is "low" (less than € 10,000; please note that the intervals are in thousands of euros), the blue line, i.e. the percentage of tax notices, is above the purple one, i.e. the percentage of coercive procedures, while for higher values of tax claim, the blue line is under the purple one. This is quite strong evidence that coercive procedures are not independent from tax claim.

As a result, the red line shows that the higher the tax claim, the higher the percentage of procedures within the tax claim range itself, up to over 70% in the last and, apparently, most desirable range.

Therefore, with just one model in place, whose task is to recognize *interesting* taxpayers, the tax authorities would risk facing many cases of coercive procedures. Thus their ability to ensure tax collection may be seriously jeopardized.

We therefore need to find a way to discover, among the most *interesting* taxpayers, the most solvent ones, the most willing to pay.

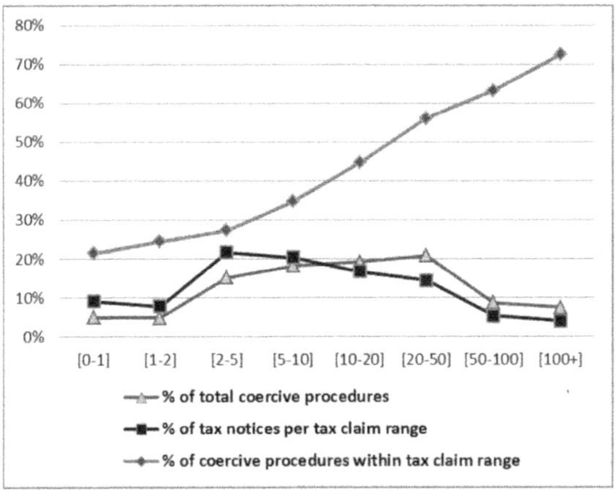

Figure 1.4. *Coercive procedures and tax claim. For a color version of this figure, see www.iste.co.uk/dimotikalis/analysis2.zip*

We can start by observing that a taxpayer with no properties will probably not be willing to pay his dues. Therefore, a second model only focusing on a few features indicating whether the taxpayer owned some kind of assets or not is built, in order to predict if a tax notice will end in an enforced recovery proceeding or not.

Once both models are available, the taxpayer selection process is held in such a way that undertakings will only be audited if judged worthy by both models.

1.2.4. *The models*

Our selection strategy needs to take into account two competing demands: on one hand, tax notices must be *profitable*, i.e. they have to address serious tax fraud or the tax evasion phenomena; on the other, *tax collectability* must be guaranteed in order to justify all of the tax authorities' efforts.

To this purpose, we develop two models, both in the form of classification trees: the first one predicts whether a taxpayer is *interesting* or not, while the second predicts the final stage of a tax notice, distinguishing between those ending with an enforced recovery proceeding and the others, where such enforced recovery proceedings do not take place.

The first one's attributes are taken from several datasets run by the IRA and are related to the taxpayers' tax returns and their annexes (such as the *sector studies*), their properties details, their customers and suppliers lists and their tax notices, whereas the second one only focuses on a set of features concerning taxpayers' assets.

In the taxpayer selection process, models that are easier to interpret are preferred to more complex models. Typically, decision trees meet the above requested conditions, so both of our models take that form.

In both cases, instead of considering just one decision tree, both practical and theoretical reasons (Breiman 1996) lead us towards a more sophisticated technique, known as *bagging*, which stands for bootstrap aggregating, with which many base classifiers are computed (in our case, many trees).

Moreover, a cost matrix is used while building the models. Indeed, in our context, to classify an actual *not interesting* taxpayer as *interesting* is a much more serious error than that of classifying as an actual *interesting* taxpayer as *not interesting*, based on the fact that, generally, tax offices' human resources are barely sufficient to perform all of the audits they are assigned. Therefore, as long as offices audit *interesting* taxpayers, everything is fine, even though many *interesting* taxpayers may not be considered. In the same way, to predict that a tax notice will not end in a coercive procedure when it actually does, is a much more serious error than that of classifying a tax notice final stage the other way round. Therefore, different weights are given to different misclassification errors.

Finally, Ross Quinlan's C4.5 decision tree algorithm is used to build the base classifiers within the *bagging* process.

Figure 1.5 puts all the pieces of our models together.

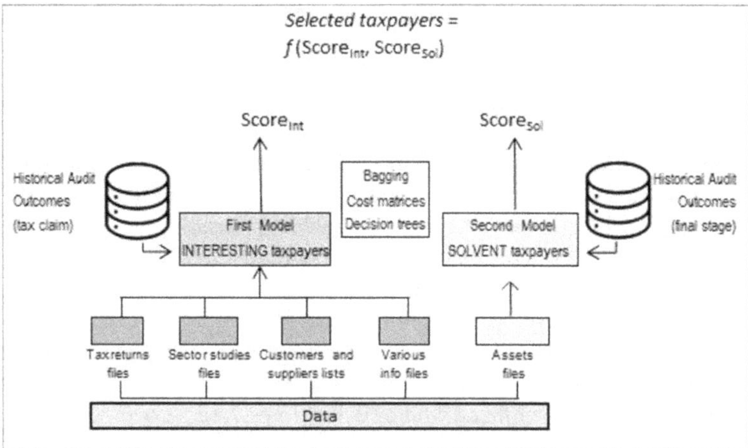

Figure 1.5. *The two models together*

1.3. Results

Our first model predicts, on the basis of the available features, 415 taxpayers to be *interesting* (i.e. 15.5% of the entire test set), with a precision rate of about 80%, as shown in Figure 1.6.

```
Correctly Classified Instances          1789               66.8535 %
Incorrectly Classified Instances         887               33.1465 %
Total Number of Instances               2676

=== Detailed Accuracy By Class ===

TP Rate  FP Rate  Precision  Recall   F-Measure  MCC      ROC Area  PRC Area  Class
0,295    0,049    0,822      0,295    0,435      0,338    0,729     0,696     interesting
0,951    0,705    0,640      0,951    0,766      0,338    0,729     0,744     not_interesting
=== Confusion Matrix ===

   a    b   <-- classified as
 341  813 |    a = interesting
  74 1448 |    b = not_interesting
```

Figure 1.6. *First model statistics and confusion matrix*

In terms of *tax claim* amounts, the model appears to perform quite well, since the selected taxpayers' average due additional taxes amounts to € 49,094, whereas the average on the entire test set is equal to € 22,339.

So far, we have shown that our model, on average, is able to distinguish serious tax evasion phenomena from the less significant ones. But what about the tax collection issue? To deal with this matter, we should investigate what kind of taxpayers we have just selected. For this purpose, Table 1.3 shows that the majority of the taxpayers, the model would select, would also be subject to coercive procedures (as we can see, the sum of the values of each column is 100%).

Pred Act	Interesting	Not Interesting
Procedure	70.12%	32.24%
No procedure	29.88%	67.76%

Table 1.3. *Predicted values versus actual coercive procedures*

Thus, many of the selected taxpayers have a debt payment issue. This jeopardizes the overall selection process efficiency and effectiveness. As pointed out by the Italian Court of Auditors, coercive procedures, on average, are able to collect only about 5% of the overall claimed credits.

To evaluate the problem extent, we can replace the actual *tax claim* value corresponding to the problematic taxpayers with the *estimated collectable tax*, which is equal to the *tax claim* multiplied by a discount factor of 95%, and compare the two scenarios, as in Figures 1.7 and 1.8, where we depict both the total *tax claim* and the average *tax claim* arising from the taxpayers' notices in the entire test set.

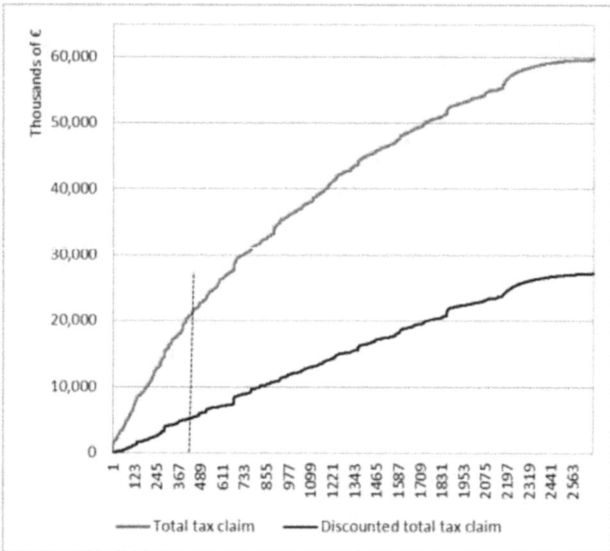

Figure 1.7. *Total tax claim and discounted tax claim. For a color version of this figure, see www.iste.co.uk/dimotikalis/analysis2.zip*

Taxpayers are ordered, from left to right, according to their probability of being *interesting*, as calculated by our model. Figure 1.7, for instance, depicts the cumulative *tax claim* charged up to a certain taxpayer: the red line values refer to the additional taxes requested with the tax notices, while the black line is drawn by considering the discounted values. The dashed vertical line indicates the levels corresponding to the last selected taxpayer according to the model (in our case, the 415th). Recall that when associating a class label with a record, the model also provides a probability, which highlights how confident the model is about its own prediction. Therefore, to a certain extent, it sets a ranking among taxpayers, which we can exploit to draw Figures 1.7 and 1.8. As we can easily observe, the overall *tax claim* charged to the selected taxpayers plummets from € 20 million to € 5 million, and the average *tax claim*, depicted in Figure 1.8, from € 49,000 to € 12,000. Thus, the selection process, which relied on our data mining model and at first sight seemed to be very efficient, shows some important flaws that we need to face. In fact, *tax collectability* is not adequately guaranteed.

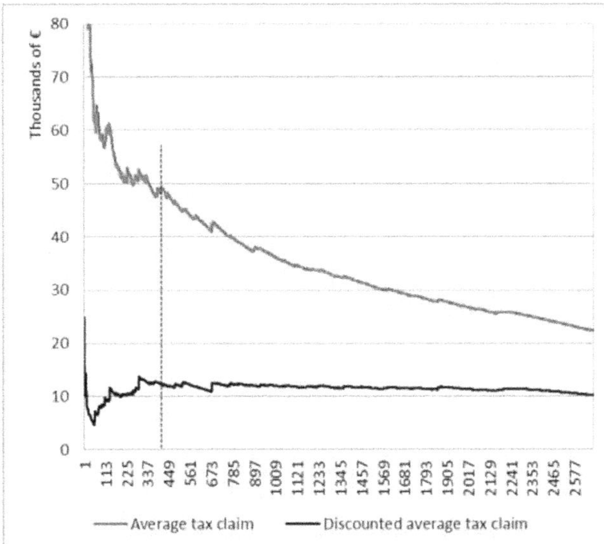

Figure 1.8. *Average total tax claim and discounted tax claim. For a color version of this figure, see www.iste.co.uk/dimotikalis/analysis2.zip*

A second model may then help us by predicting which taxpayers would not be subject to coercive procedures, by focusing on a set of features concerning their assets.

Again, with a precision rate of about 80%, as shown in Figure 1.9, the model appears to be successful.

```
Correctly Classified Instances        1480               55.3064 %
Incorrectly Classified Instances      1196               44.6936 %
Total Number of Instances             2676

=== Detailed Accuracy By Class ===

TP Rate  FP Rate  Precision  Recall   F-Measure  MCC      ROC Area  PRC Area  Class
0,874    0,644    0,455      0,874    0,598      0,251    0,688     0,532     Procedure
0,356    0,126    0,820      0,356    0,496      0,251    0,688     0,764     NO_procedure
=== Confusion Matrix ===

   a    b   <-- classified as
  891  129 |    a = Procedure
 1067  589 |    b = NO_procedure
```

Figure 1.9. *Second model statistics and confusion matrix*

Pred / Act	Procedure	No Procedure
Interesting	46.94%	32.73%
Not interesting	53.06%	67.27%

Table 1.4. *Predicted coercive procedures versus actual interesting taxpayers*

This second model could be useful on our end, even though it may have some *caveats*. First, most of the taxpayers that the model classifies as people that will not face a coercive procedure are also *not interesting*, as shown in Table 1.4. Again, the sum of the values of each column is 100%.

In fact, this second model's performance in terms of *tax claim* appears to have worsened with respect to the first, since the *no procedure* taxpayers' average due additional tax, calculated on the first 415 taxpayers (according to the ranking set by this model, which is, obviously, dramatically different from the one set by the first model we have seen), is equal to € 20,388. However, the average *collectable tax claim* is equal to € 13,493, which is a little bit better than the one we have seen before.

We point out that throughout this chapter, we have compared sets of selected taxpayers with the same cardinality, for two kinds of considerations: first, tax authorities, reasonably, have a fixed budget of audits to perform, so comparisons between models should be done subject to a given number of audits; second, for comparability reasons, since smaller sets tend to perform more (see Figure 1.8, where the average *tax claim* decreases while the number of selected taxpayers increases).

Therefore, in this second model we have developed, the high rate of *not interesting taxpayers*, on one hand, causes a drop in the average *tax claim* (from 49,000 to 20,000), but, on the other, it contributes to the slight enhancement of the discounted average *tax claim* (from € 12,000 to € 13,000), since only a few of the *not interesting* taxpayers pass through a coercive procedure. Figure 1.10 compares, for each number of selected taxpayers, the different coercive procedures rates arising from the two models.

What we can do, then, is use the two models "together". For instance, we could exploit the first model in order to sort the taxpayers eligible to be selected and the second one to discard the ones likely to be subject to coercive procedures.

In such a way, if we imagine that we select our 415 taxpayers again, on one hand, we would select both *interesting* and *not interesting* taxpayers (only if the second model had predicted that no *interesting* taxpayers would go through a coercive procedure, we would have selected only *interesting* taxpayers), but, on the other, we would also select the taxpayers who are more likely to pay their tax debts.

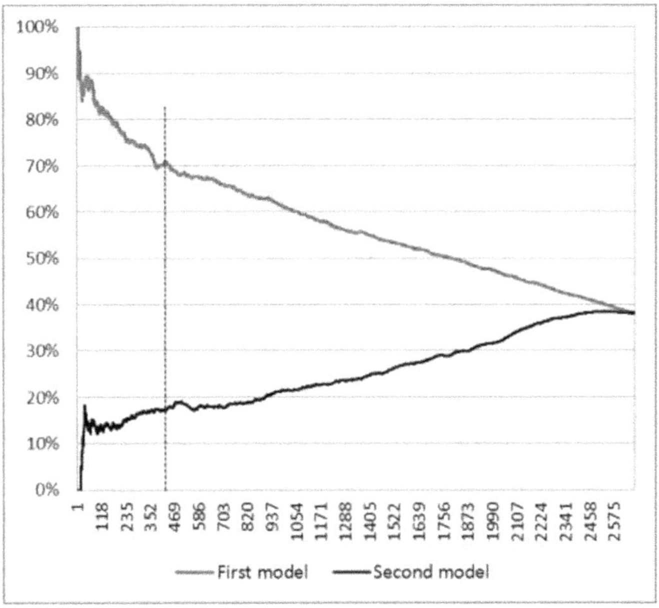

Figure 1.10. *Coercive procedures' rates. For a color version of this figure, see www.iste.co.uk/dimotikalis/analysis2.zip*

This is just an example and it is not the only way we can combine the two models. Indeed, there is space for policymakers to exploit the two models in different ways, depending on the kind of tradeoff choices they may want to reach, concerning the two goals of the audit process: its profitability and its tax collectability. For instance, a selection process could only be targeted towards interesting taxpayers *and* taxpayers without payment issues.

Anyway, does the tradeoff we have sketched above work?

Figures 1.11–1.13 can shed some light on our *ensemble model*'s performance. As usual, the dashed vertical line shows the values corresponding to the number of taxpayers we wish to select.

In our case, thus, with the *ensemble model*, we would claim, on average, € 26,219 from the selected taxpayers and we would hopefully collect, on average, € 17,542 from each of them, of whom only 25% are predicted to incur in coercive procedures.

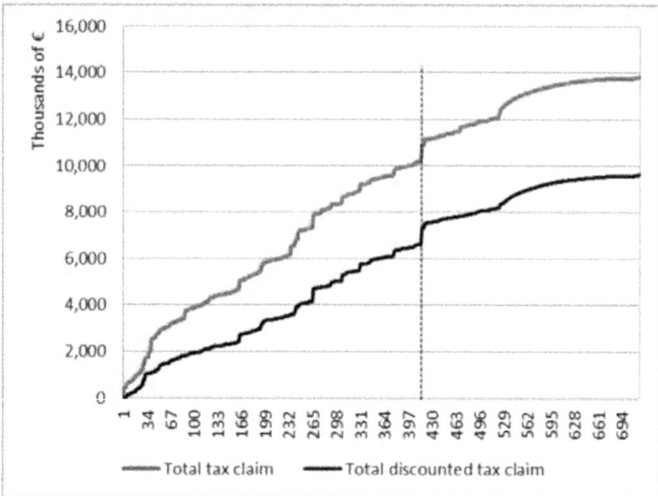

Figure 1.11. *Total tax claim. For a color version of this figure, see www.iste.co.uk/dimotikalis/analysis2.zip*

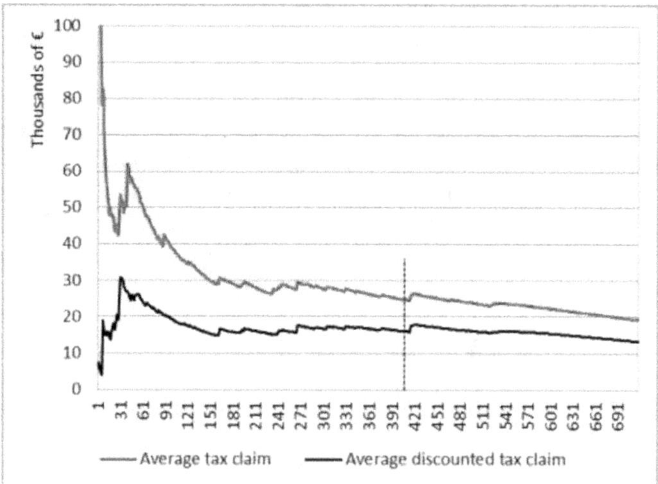

Figure 1.12. *Average tax claim. For a color version of this figure, see www.iste.co.uk/dimotikalis/analysis2.zip*

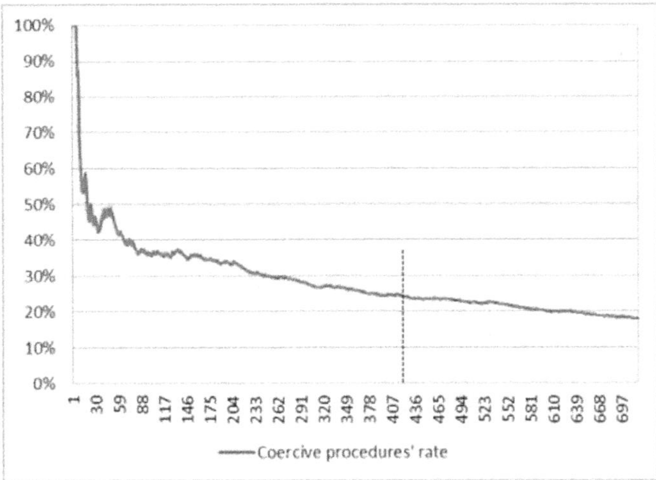

Figure 1.13. *Coercive procedures' rate. For a color version of this figure, see www.iste.co.uk/dimotikalis/analysis2.zip*

In a hypothetical selection process, the winning strategy would then be to use the *ensemble model*, since it maximizes the *collectable tax claim*.

What we might be interested in, is to know whether the *ensemble model* is always the best option. This may depend on the coercive procedures' rate that characterizes the two sets of auditable taxpayers selected by the two models. Unfortunately, once we build the models, before applying them to the test set, we do not exactly know what kind of taxpayers will be selected. Therefore, we do not even know these rates; however, we can consider them as unknown parameters, say θ' and θ''. From this point of view, the rates we have observed within the two selected sets can be considered as two values of such parameters, say $\widehat{\theta'}$ (70%) and $\widehat{\theta''}$ (25%) (see Table 1.5).

To satisfy our interest, we should depict the two models' behavior as a function of the unknown parameters, θ' and θ'', respectively; that is, we should calculate the expected tax revenues amounts for any value of θ' and θ''. Unfortunately, this cannot be done. To understand why, suppose that for both models, only one of the selected taxpayers turns out to be subject to coercive procedures. If this taxpayer's debt is high, the amount of money that is difficult to collect would be high, but if his debt is low, then the uncollected tax would also be low.

What can be done, instead, is to calculate, for any given value of θ' and θ'', the *maximum* and *minimum* collectable taxes arising from each model. Indeed, the *maximum* collectable taxes scenario is the one where coercive procedures are first applied to the less unfaithful taxpayers, while the *minimum* collectable taxes scenario refers to a situation in which coercive procedures are first applied to the most unfaithful taxpayers, as shown in Figure 1.14.

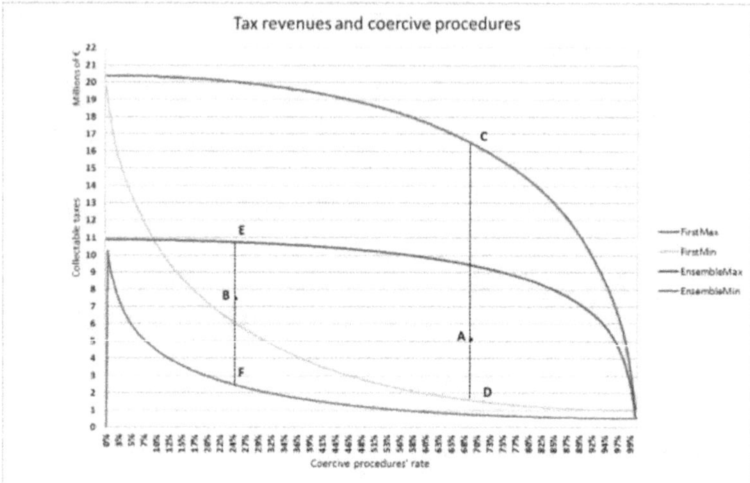

Figure 1.14. *Models' maximum and minimum collected tax. For a color version of this figure, see www.iste.co.uk/dimotikalis/analysis2.zip*

The first model's *maximum* and *minimum* values are represented by the red and orange lines, while the *ensemble model*'s are the blue and purple ones. Any point within the red and orange lines represents a possible outcome for the first model and any point within the blue and purple lines represents one possible outcome for the *ensemble model*. For instance, points A and B represent the outcomes of our models (the first and the *ensemble*, respectively), given our training and test sets.

Having to deal with two areas means that the models' behavior is determined not only by θ' and θ'', but also by the kind of taxpayers that go through a coercive procedure. If we could shrink the areas between the red and orange lines and the blue and purple ones, we could be put in a better shape.

How could we do this? Well, if we turn back to points A and B in Figure 1.14, and we draw two dashed vertical lines from them, we can see that the first is nearer to the *minimum* line of its model (since line $\overline{AD}$ is shorter than line $\overline{CA}$), while the other is nearer to the *maximum* one (since line $\overline{EB}$ is shorter than line $\overline{BF}$).

If we assume that, for each value of θ' and θ'' and for each corresponding points A (and, also, lines $\overline{AD}$ and $\overline{CA}$) and B (and, also, lines $\overline{EB}$ and $\overline{BF}$), ratios $\frac{\overline{CA}}{\overline{AD}}$

are always the same and also ratios $\frac{\overline{BF}}{\overline{EB}}$, we could draw a single line for each model, which would only be a function of θ' and θ''', respectively, as shown in Figure 1.15.

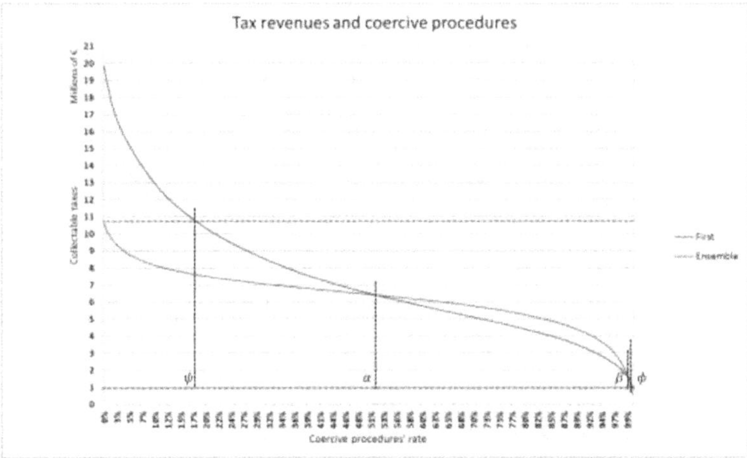

Figure 1.15. *Models' approximation. For a color version of this figure, see www.iste.co.uk/dimotikalis/analysis2.zip*

Therefore, we have to study the two monotonically decreasing functions, say $\gamma_{first}(\theta')$ and $\gamma_{ens}(\theta'')$ to find out for which joint values of θ' and θ'', one model is better than the other.

Based on our data, these functions intersect at two points, where θ' and θ'' are, respectively, equal to α and β. Moreover:

– $\gamma_{first}(0) > \gamma_{ens}(0)$, i.e. if all taxpayers were to pay their debts, the first model would be better than the ensemble one.

– $\gamma_{first}(1) > \gamma_{ens}(1)$ since if all taxpayers were to undergo a coercive procedure, these functions' values would be 0.05 times $\gamma_{first}(0)$ and $\gamma_{ens}(0)$, respectively (recall that in the case of coercive procedures, the *collectable tax* is assumed to be equal to the *tax claim* multiplied by a discount factor of 95%).

– $\gamma_{first}(\theta') \geq \gamma_{ens}(\theta'')$, for $\theta' \leq \alpha$ and $\theta'' \geq \alpha$

– $\gamma_{first}(\theta') < \gamma_{ens}(\theta'')$, for $\alpha < \theta' < \phi$ and $\theta'' < \alpha$

– $\gamma_{first}(\theta') \geq \gamma_{ens}(\theta'')$, for $\alpha \leq \theta' \leq \beta$ and $\theta'' \geq \beta$

– $\gamma_{first}(\theta') < \gamma_{ens}(\theta'')$, for $\beta < \theta' < \phi$ and $\alpha < \theta'' < \beta$

– $\gamma_{first}(\theta') > \gamma_{ens}(\theta'')$, for $\beta < \theta' < \phi$ and $\theta'' > \phi$

– There is a ψ such that $\gamma_{first}(\theta') \geq \gamma_{ens}(\theta'')$, for $\theta' \leq \psi$ and for any θ''.

– There is a ϕ such that $\gamma_{first}(\theta') \geq \gamma_{ens}(\theta'')$, for $\theta' \geq \phi$ and for any θ''.

Figure 1.16 depicts, in a θ' x θ'' space, the regions where the two models represent the best choice (the dark gray region is where the first model is the best option, while in the light gray one, the *ensemble model* is better).

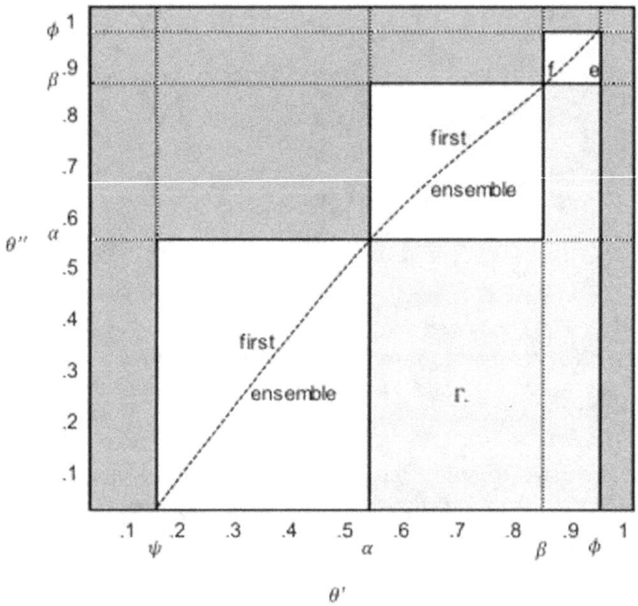

Figure 1.16. *Values of θ' and θ'' determining the best model. For a color version of this figure, see www.iste.co.uk/dimotikalis/analysis2.zip*

In the three white regions, the exact combinations of θ' and θ'' that guarantee whether a model is better than the other, depend on the relative slopes of $\gamma_{first}(\theta')$ and $\gamma_{ens}(\theta'')$.

As a general rule, if we expect small values of θ' or high values of θ', and also high values of θ'' in our samples of auditable taxpayers, then the first model is likely to guarantee a higher revenue; otherwise, the *ensemble model* is the one that we should use. From Figure 1.16, we note that our experience on the 8,000 taxpayers dataset took us to point Γ, which lies in a region where the *ensemble model* is the best option.

	First model	Second model	Ensemble model	Test set
Number of selected taxpayers	415	415	415	2,676
Interesting taxpayers rate	82.20%	32.77%	42.89%	43.12%
Coercive procedures rate	70.12%	17.35%	24.58%	38.12%
Average tax claim (€)	49,094	20,388	26,219	22,339
Average collectable tax (€)	12,187	13,493	17,542	10,194

Table 1.5. *The most significant results of the models*

1.4. Discussion

The learning scheme developed in this chapter is aimed at computing a risk factor for each taxpayer, optimizing the tax authorities' audit processes, taking into account two competing needs: the profitability of each tax notice and the effective collectability of the additional requested taxes.

The *ensemble model* seems to tackle both of the above-mentioned issues quite well.

Given that the whole test set's average claim is € 22,339, while the average collectable taxes are equal to € 10,194, our procedure increases the first figure by 1.17% (€ 26,219) and the second by 72% (€ 17,542).

With respect to the scenario in which only the first model is put in place, by developing the twofold selection process as described above, the presence of coercive procedures dramatically plummets from 70% to 25%. Moreover, the selection of *not interesting* taxpayers, while causing a drop in the average *tax claim* (from € 49,094 to € 26,219), is more than compensated by the procedure's capability of efficiently collecting the additional taxes charged to the selected taxpayers (from € 12,187 to € 17,542).

Table 1.5 summarizes the most significant results reached by the three models that have been built: the first model looks for *interesting* taxpayers; the second model is in search of solvent taxpayers; and the third model, called the *ensemble model*, is a combination of the first two. To better understand the figures referred to the models, the same information set is shown, related to the entire test set.

This result can be generalized, and the best selection strategy depends on our estimates of θ' and θ'' in the sets of the selected taxpayers.

1.5. Conclusion

The data analysis framework designed in this chapter gives an effective learning scheme aimed at improving the IRA's ability to identify non-compliant taxpayers.

It involves two C4.5 decision trees, predicting two different class values, based on two different predictive attribute sets. That is, the first model is built to identify the most likely non-compliant taxpayers, while the second one identifies the ones who are more likely going to pay the additional tax bill. This twofold selection process target is requested in order to maximize the overall audit effectiveness, so businesses will be audited, only if suggested by both models.

Tax evasion is a topic that has been studied extensively in the past (starting from Allingham and Sandmo 1972) and it is still a hot topic. Most models are usually mainly concerned with finding the best way to identify the most relevant cases of tax evasion. In this chapter, we go further, analyzing the overall effectiveness of the tax authorities activity, which has to take into account both the tax notices' profitability and the collectability of the additional requested taxes.

The latter issue cannot be tackled without knowing the final stage of the tax notices. In fact, it is very difficult to have this kind of information at hand, even because a tax notice can come to an end years after it was sent to the taxpayer (especially when a tax court is addressed).

By ignoring the collectability aspect of the audit process, the selection processes may not be correctly targeted, or at least, may not satisfy the tax authorities' needs i.e. relevant evasion phenomena may be discovered, but only little money may be collected.

Of course, the fight against tax evasion is not only a matter of collecting money, but should also have some other purposes, such as promoting taxpayers' compliant behavior. Nonetheless, efficient tax bill collection is crucial from the state budget point of view, because public expenditures are strictly connected to public revenues.

The methodology we suggest here will soon be validated in real cases i.e. a sample of taxpayers will be selected according to the classification criteria developed in this chapter and will subsequently be involved in some audit processes.

1.6. References

Agenzia delle Entrate e Ministero dell'Economia e delle Finanze (2018). Convenzione triennale per gli esercizi 2018–2020 [Online]. Available at: https://www.finanze.it/export/sites/finanze/.galleries/Documenti/Varie/DF_CONVENZIONE-MEF_ADE_2018.2020_FIRMATA-28_11_2018.pdf.

Allingham, M.G. and Sandmo, A. (1972). Income tax evasion: A theoretical analysis. *Journal of Public Economics*, I, 323–338.

Barone, M., Pisani, S., Spingola, A. (2017). Data mining application issues in income indicators audits. *Argomenti di discussione – Agenzia delle Entrate*, 2.

Basta, S., Fassetti, F., Guarascio, M., Manco, G., Giannotti, F., Pedreschi, D., Spinsanti, L., Papi, G., Pisani, S. (2009). High quality true positive prediction for fiscal fraud detection to regressive conditional. *2009 IEEE International Conference on Data Mining Workshops*.

Bonchi, F., Giannotti, F., Mainetto, G., Pedreschi, D. (1999). A classification-based methodology for planning auditing strategies in fraud detection. *Proc. of SIGKDD99*, 175–184.

Breiman, L. (1996). Bagging predictors. *Machine Learning*, 24(2), 123–140.

Corte dei Conti (2016). Il sistema della riscossione dei tributi erariali al 2015. *Deliberazione 20 ottobre 2016*, 11/2016/G.

Gonzalez, P.C. and Velasquez, J.D. (2013). Characterization and detection of taxpayers with false invoices using data mining techniques. *Expert Systems with Applications*, 40(5), 1427–1436.

OpenStax (2013). Introductory Statistics. OpenStax, 19 September [Online]. Available at: http://cnx.org/content/col11562/latest/.

Phua, C., Lee, V., Smith, K., Gayler, R. (2005). A comprehensive survey of data mining-based fraud detection research. *Artificial Intelligence Review*, submitted.

de Roux, D., Perez, B., Moreno, A., del Pilar Villamil, M., Figueroa, C. (2018). Tax fraud detection for under-reporting declarations using an unsupervised machine learning approach. *KDD 2018*, 215–222.

de Sisti, P. and Pisani, S. (2007). Data mining e analisi del rischio di frode fiscale: il caso dei crediti d'imposta. *Documenti di lavoro dell'Ufficio Studi – Agenzia delle Entrate*, 4.

Wu, R., Ou, C.S., Lin, H., Chang, S., Yen, D. (2012). Using data mining technique to enhance tax evasion detection performance. *Expert Systems with Applications*, 39, 8769–8777.

2

Asymptotics of Implied Volatility in the Gatheral Double Stochastic Volatility Model

Gatheral's (2008) double-mean-reverting model by is motivated by empirical dynamics of the variance of stock price. No closed-form solution for European option exists in the above model. In this chapter, we study the behavior of the implied volatility with respect to the logarithmic strike price and maturity near expiry and at-the-money. Using the method by Pagliarani and Pascucci (2017), we explicitly calculate the first few terms of the asymptotic expansion of the implied volatility within a parabolic region.

2.1. Introduction

The history of implied volatility can be traced back at least to Latané and Rendleman (1976), where it appeared under the name "*implied standard deviation*", i.e. the standard deviation of asset returns, which are *implied* in actual European call option prices when investors price options according to the Black–Scholes model. For a recent review of different approaches to determine implied volatility, see Orlando and Taglialatela (2017). To give exact definitions, we use Pagliarani and Pascucci (2017).

In order to briefly explain our contribution to the subject, we will introduce some notations. Let $d \geq 2$ be a positive integer, let $T_0 > 0$ be a time horizon, let $T \in (0, T_0]$, and let $\{ \mathbf{Z}_t : 0 \leq t \leq T \}$ be a continuous $\mathbb{R}^d$-valued adapted Markov stochastic process on a probability space $(\Omega, \mathfrak{F}, \mathsf{P})$ with a filtration $\{ \mathfrak{F}_t : 0 \leq t \leq T \}$. Assume that the first

Chapter written by Mohammed ALBUHAYRI, Anatoliy MALYARENKO, Sergei SILVESTROV, Ying NI, Christopher ENGSTRÖM, Finnan TEWOLDE and Jiahui ZHANG.

coordinate S_t of the process $\mathbf{Z}_t$ represents the risk-neutral price of a financial asset, and the $d-1$ remaining coordinates $\mathbf{Y}_t$ represent stochastic factors in a market with zero interest rate and no dividends.

On one hand, we have the time t no-arbitrage price of a European call option with strike price $K>0$ and maturity T is $C_{t,T,K}=v(t,S_t,\mathbf{Y}_t,T,K)$, where

$$v(t,s,\mathbf{y},T,K)=\mathsf{E}[\max\{0,S_T-K\}\,|\,\mathfrak{F}_t,S_t=s,\mathbf{Y}_t=\mathbf{y}],$$

and where $(t,s,\mathbf{y})\in[0,T]\times(0,\infty)\times\mathbb{R}^{d-1}$. We change to *logarithmic variables* and define the option price by

$$u(t,x,\mathbf{y},T,k)=v(t,\mathrm{e}^x,\mathbf{y},T,\mathrm{e}^k),$$

where x is the time t log price of the underlying asset, k is the log strike of the option, and $(t,x,\mathbf{y})\in[0,T]\times\mathbb{R}\times\mathbb{R}^{d-1}$.

On the other, the Black–Scholes price in logarithmic variables is

$$u^{\mathrm{BS}}(\sigma,\tau,x,k)=\mathrm{e}^x\mathcal{N}(d_+)-\mathrm{e}^k\mathcal{N}(d_-),\quad \text{where}\quad d_\pm=\frac{1}{\sigma\sqrt{\tau}}\left(x-k\pm\frac{\sigma^2\tau}{2}\right), \qquad [2.1]$$

and $\tau=T-t\in[0,T]$, x, $k\in\mathbb{R}$, $\mathcal{N}$ is the cumulative distribution function of the standard normal random variable.

DEFINITION 2.1.– *The implied volatility $\sigma=\sigma(t,x,\mathbf{y},T,k)$ is the unique positive solution of the nonlinear equation*

$$u^{\mathrm{BS}}(\sigma,\tau,x,k)=u(t,x,\mathbf{y},T,k).$$

REMARK 2.1.– *In the literature on option pricing, there are concepts of model implied volatility and market implied volatility. If the right-hand side of the above equation, i.e. $u(t,x,\mathbf{y},T,k)$, refers to the European option price under a given model, then $\sigma=\sigma(t,x,\mathbf{y},T,k)$ is called the model implied volatility. If $u(t,x,\mathbf{y},T,k)$ is replaced by the observed market option price, then we have the so-called market implied volatility. Here, we work with the (model) implied volatility.*

Pagliarani and Pascucci (2012) derived a fully explicit approximation for the implied volatility at any given order $N\geq 0$ for the scalar case. Lorig *et al.* (2017) extended this result to the multidimensional case. Denote the above approximation by $\overline{\sigma}_N(t,x,\mathbf{y},T,k)$.

Pagliarani and Pascucci (2017) proved that under some mild conditions, the following limits exist:

$$\frac{\partial^q}{\partial T^q}\frac{\partial^m}{\partial k^m}\overline{\sigma}_N(t,x) = \lim_{(T,k)\to(t,x)} \frac{\partial^q}{\partial T^q}\frac{\partial^m}{\partial k^m}\overline{\sigma}_N(t,x,T,k),$$

where the limit is taken as (T,k) approaches (t,x) within the parabolic region

$$\mathscr{P}_\lambda = \{(T,k) \in (0,T_0] \times \mathbb{R} \colon |x-k| \leq \lambda\sqrt{T-t}\}$$

for an arbitrary positive real number λ and nonnegative integers m and q.

Moreover, Pagliarani and Pascucci (2017) established an asymptotic expansion of the implied volatility in the following form:

$$\begin{aligned}\sigma(t,x,\mathbf{y},T,k) = & \sum_{2q+m\leq N} \frac{1}{q!m!}\frac{\partial^q}{\partial T^q}\frac{\partial^m}{\partial k^m}\overline{\sigma}_N(t,x)(T-t)^q(k-x)^m \\ & + o((T-t)^{N/2} + |k-x|^N),\end{aligned} \tag{2.2}$$

as (T,k) approaches (t,x) within $\mathscr{P}_\lambda$.

We apply the above described theory to the double-mean-reverting model by Gatheral (2008) given by the following system of stochastic differential equations:

$$\begin{aligned} \mathrm{d}S_t &= \sqrt{v_t}S_t\,\mathrm{d}W_t^1, \\ \mathrm{d}v_t &= \kappa_1(v_t' - v_t)\,\mathrm{d}t + \xi_1 v_t^{\alpha_1}\,\mathrm{d}W_t^2, \\ \mathrm{d}v_t' &= \kappa_2(\theta - v_t')\,\mathrm{d}t + \xi_2 v_t'^{\alpha_2}\,\mathrm{d}W_t^3, \end{aligned} \tag{2.3}$$

where the Wiener processes W_t^i are correlated: $\mathsf{E}[W_s^i W_t^j] = \rho_{ij}\min\{s,t\}$, and where parameters $\kappa_1, \kappa_2, \theta, \xi_1, \xi_2, \alpha_1, \alpha_2$ are the positive real numbers. Note that while S_0 is observable in the market, v_0 and v_0' are usually not observable and may be calibrated from the market data on options.

In this model, with rate κ_1 the variance v_t mean reverts to a level v_t' which itself moves over time to the level θ at a (usually slower rate) κ_2, hence the name double-mean-reverting. Here, parameters $\alpha_1, \alpha_2 \in [1/2, 1]$. In the case of $\alpha_1 = \alpha_2 = 1/2$, we have the so-called double Heston model; in the case of $\alpha_1 = \alpha_2 = 1$, the double lognormal model; and finally, in the general case, the double CEV model (Gatheral 2008).

The DMR model can be consistently calibrated to both the SPX options and the VIX options. However, due to the lack of an explicit formula for both the European option price and the implied volatility, the calibration is usually done using time-consuming methods like the Monte Carlo simulation or the finite difference method. In this chapter, we provide an explicit solution to the implied volatility under this model.

In section 2.2, we formulate three theorems that give the asymptotic expansions of implied volatility of orders 0, 1 and 2. Detailed proof of Theorems 2.1 and 2.2 as well as a short proof of Theorem 2.3 without technicalities are given in section 2.3.

2.2. The results

Put $x_t = \ln S_t$.

THEOREM 2.1.– *The asymptotic expansion of order* 0 *of the implied volatility has the form*

$$\sigma(t,T) = \sqrt{v_0} + o(1).$$

THEOREM 2.2.– *The asymptotic expansion of order* 1 *of the implied volatility has the form*

$$\sigma(t,x_0,v_0,v_0';T,k) = \sqrt{v_0} + \frac{1}{4}\rho_{12}\xi_1 v_0^{\alpha_1-1}(k-x_0) + o(\sqrt{T-t} + |k-x_0|).$$

THEOREM 2.3.– *The asymptotic expansion of order* 2 *of the implied volatility has the form*

$$\begin{aligned}\sigma(t,x_0,v_0,v_0';T,k) = {} & \sqrt{v_0} + \frac{1}{4}\rho_{12}\xi_1 v_0^{\alpha_1-1}(k-x_0) \\ & - \frac{3}{16}\rho_{12}^2\xi_1^2 v_0^{2\alpha_1-5/2}(k-x_0)^2 \\ & + \frac{1}{32}[8\kappa_1(v_0'-v_0)/\sqrt{v_0} + 2\rho_{12}\xi_1 v_0^{\alpha_1} + 3\rho_{12}^2\xi_1^2 v_0^{2\alpha_1-3/2}](T-t) \\ & + o(T-t+(k-x_0)^2).\end{aligned} \qquad [2.4]$$

2.3. Proofs

PROOF OF THEOREM 2.1.– First, we perform the change of variable $x_t = \ln S_t$ in the system [2.3]. Using the multidimensional Itô formula, we obtain

$$\begin{aligned} dx_t &= -\frac{1}{2}v_t\,dt + \sqrt{v_t}\,dW_t^1, \\ dv_t &= \kappa_1(v_t'-v_t)\,dt + \xi_1 v_t^{\alpha_1}\,dW_t^2, \\ dv_t' &= \kappa_2(\theta - v_t')\,dt + \xi_2 v_t'^{\alpha_2}\,dW_t^3. \end{aligned}$$

The infinitesimal generator of this system is

$$\mathscr{A} = \sum_{i,j=1}^{3} a_{ij}(\mathbf{z}) \frac{\partial^2}{\partial z_i \partial z_j} + \sum_{i=1}^{3} a_i(\mathbf{z}) \frac{\partial}{\partial z_i}$$

with $\mathbf{z} = (x, y, z)^\top$. We have

$$a_{11}(\mathbf{z}) = \frac{1}{2}y, \qquad a_{12}(\mathbf{z}) = \frac{1}{2}\rho_{12}\xi_1 y^{\alpha_1+1/2}, \qquad a_{13}(\mathbf{z}) = \frac{1}{2}\rho_{13}\xi_2\sqrt{y}z^{\alpha_2},$$

$$a_{22}(\mathbf{z}) = \frac{1}{2}\xi_1^2 y^{2\alpha_1}, \quad a_{23}(\mathbf{z}) = \frac{1}{2}\rho_{23}\xi_1\xi_2 y^{\alpha_1} z^{\alpha_2}, \quad a_{33}(\mathbf{z}) = \frac{1}{2}\xi_2^2 z^{2\alpha_2},$$

$$a_1(\mathbf{z}) = -\frac{1}{2}y, \qquad a_2(\mathbf{z}) = \kappa_1(z-y), \qquad a_3(\mathbf{z}) = \kappa_2(\theta - z).$$

From Pagliarani and Pascucci (2017), Definition 3.4, we have

$$\overline{\sigma}_N(t,x,y,z;T,k) = \sum_{n=0}^{N} \sigma_n^{(x,y,z)}(t,x,y,z;T,k), \qquad [2.5]$$

where the terms on the right-hand side of equation [2.5] are the values of the functions $\sigma_n^{(\bar{x},\bar{y},\bar{z})}(t,x,y,z;T,k)$ given by (Pagliarani and Pascucci 2017, Equation 3.15), when $\bar{x} = x$, $\bar{y} = y$, and $\bar{z} = z$. (Pagliarani and Pascucci 2017, Equation 3.15) is recursive, and we define the above functions for $n = 0$ first.

Following Lorig *et al.* (2017), Appendix B, put $\bar{x} = x_0$, $\bar{y} = v_0$, $\bar{z} = v_0'$. From Lorig *et al.* (2017), Equation 3.2, we have

$$\sigma_0^{(\bar{\mathbf{z}})}(t,x,y,z;T,k) = \sqrt{2a_{11}(\bar{\mathbf{z}})},$$

where $\bar{\mathbf{z}} = (\bar{x},\bar{y},\bar{z})^\top$. It follows that $\sigma_0^{(\bar{\mathbf{z}})}(t,x,y,z;T,k) = \sqrt{\bar{y}}$. Then, we have

$$\overline{\sigma}_0(t,x_0,y_0,z_0;T,k) = \sqrt{\bar{y}} = \sqrt{v_0}$$

and Theorem 2.1 follows from [2.2] and [2.5].

PROOF OF THEOREM 2.2.– Let $n \geq 1$, and let h be an integer with $1 \leq h \leq n$. The *Bell polynomials* are defined by Pagliarani and Pascucci (2017) in Equation E.5

$$B_{n,h}(z_1,z_2,\dots,z_{n-h+1}) = n! \sum_{j_1,j_2,\dots,j_{n-h+1}} \prod_{i=1}^{n-h+1} \frac{1}{j_i!}\left(\frac{z_i}{i!}\right)^{j_i},$$

where the sum is taken over all sequences $\{\, j_i \colon 1 \leq i \leq n-h+1\}$ of non-negative integers such that

$$\sum_{i=1}^{n-h+1} j_i = h, \qquad \sum_{i=1}^{n-h+1} i j_i = n.$$

Let $u^{\mathrm{BS}}(\sigma;\tau,x,k)$ be the Black–Scholes price [2.1]. Pagliarani and Pascucci (2017, Equation 3.15) has the form

$$\begin{aligned}\sigma_n^{(\bar{\mathbf{z}})}(t,x,y,z;T,k) &= u_n^{(\bar{\mathbf{z}})}\left(\frac{\partial}{\partial\sigma}u^{\mathrm{BS}}\left(\sigma_0^{(\bar{\mathbf{z}})}\right)\right)^{-1}\\ &\quad -\frac{1}{n!}\sum_{h=2}^{N}B_{n,h}(1!\sigma_1^{(\bar{\mathbf{z}})},2!\sigma_2^{(\bar{\mathbf{z}})},\dots,(n-h+1)!\sigma_{n-h+1}^{(\bar{\mathbf{z}})}) \qquad [2.6]\\ &\quad \times\frac{\partial^h}{\partial\sigma^h}u^{\mathrm{BS}}\left(\sigma_0^{(\bar{\mathbf{z}})}\right)\left(\frac{\partial}{\partial\sigma}u^{\mathrm{BS}}\left(\sigma_0^{(\bar{\mathbf{z}})}\right)\right)^{-1}, \qquad n\geq 1.\end{aligned}$$

For the sake of simplicity, we have omitted the last three arguments of the function u^{BS} and all arguments of the functions $\sigma_i^{(\bar{\mathbf{z}})}$, $0\leq i\leq n$ and $u_n^{(\bar{\mathbf{z}})}$.

To define $u_n^{(\bar{\mathbf{z}})}$, consider the differential operator

$$\mathscr{A}_n^{(\bar{\mathbf{z}})}(\mathbf{z}) = \sum_{i,j=1}^{3} a_{ij,n}(\mathbf{z})\frac{\partial^2}{\partial z_i\partial z_j} + \sum_{i=1}^{3} a_{i,n}(\mathbf{z})\frac{\partial}{\partial z_i},$$

where

$$a_{ij,n}(\mathbf{z}) = \sum_{\beta_1+\beta_2+\beta_3=n}\frac{\partial^n a_{ij}}{\partial x^{\beta_1}\partial y^{\beta_2}\partial z^{\beta_3}}(\bar{\mathbf{z}})\frac{(x-\bar{x})^{\beta_1}(y-v_0)^{\beta_2}(z-v_0')^{\beta_3}}{\beta_1!\beta_2!\beta_3!} \qquad [2.7]$$

and

$$a_{i,n}(\mathbf{z}) = \sum_{\beta_1+\beta_2+\beta_3=n}\frac{\partial^n a_{i}}{\partial x^{\beta_1}\partial y^{\beta_2}\partial z^{\beta_3}}(\bar{\mathbf{z}})\frac{(x-\bar{x})^{\beta_1}(y-v_0)^{\beta_2}(z-v_0')^{\beta_3}}{\beta_1!\beta_2!\beta_3!}$$

are the terms of the Taylor expansions of the functions $a_{ij}(\mathbf{z})$ and $a_i(\mathbf{z})$ around the point $\bar{\mathbf{z}}$.

Following Pagliarani and Pascucci (2017), define the vector $\mathbf{m}^{(\bar{\mathbf{z}})}(t,s)$ by

$$m_i^{(\bar{\mathbf{z}})}(t,s) = (s-t)a_i(\bar{\mathbf{z}}), \qquad 1\leq i\leq 3,$$

the matrix $C^{(\bar{\mathbf{z}})}(t,s)$ by

$$C_{ij}^{(\bar{\mathbf{z}})}(t,s) = (s-t)a_{ij}(\bar{\mathbf{z}}), \qquad 1\leq i,j\leq 3,$$

and the operator $\mathscr{G}_n^{(\bar{\mathbf{z}})}(t,s,\mathbf{z})$ by

$$\mathscr{G}_n^{(\bar{\mathbf{z}})}(t,s,\mathbf{z}) = \mathscr{A}_n^{(\bar{\mathbf{z}})}(\mathbf{z}-\bar{\mathbf{z}}+\mathbf{m}^{(\bar{\mathbf{z}})}(t,s)+C^{(\bar{\mathbf{z}})}(t,s)\nabla_{\mathbf{z}}). \qquad [2.8]$$

Define the set $I_{n,h}$ by

$$I_{n,h} = \{\mathbf{i} = (i_1, \dots, i_h) \colon i_1 + \cdots + i_h = n\},$$

and the operator $\mathscr{L}_n^{(\bar{\mathbf{z}})}(t,T,\mathbf{z})$ as the differential operator acting on the $\mathbf{z}$-variable and defined by (Pagliarani and Pascucci 2017, Equation D.2) as

$$\mathscr{L}_n^{(\bar{\mathbf{z}})}(t,T,\mathbf{z}) = \sum_{h=1}^{n} \int_t^T \int_{s_1}^T \cdots \int_{s_{h-1}}^T \sum_{\mathbf{i} \in I_{n,h}} \mathscr{G}_{i_1}^{(\bar{\mathbf{z}})}(t,s_1,\mathbf{z}) \cdots \mathscr{G}_{i_h}^{(\bar{\mathbf{z}})}(t,s_h,\mathbf{z})\, \mathrm{d}s_h \cdots \mathrm{d}s_1.$$

The function $u_n^{(\bar{\mathbf{z}})}$ in equation [2.6] is defined by (Pagliarani and Pascucci 2017, Equation D.1)

$$u_n^{(\bar{\mathbf{z}})}(t,\mathbf{z};T,k) = \mathscr{L}_n^{(\bar{\mathbf{z}})}(t,T,\mathbf{z}) u^{\mathrm{BS}}\left(\sigma_0^{(\bar{\mathbf{z}})};\tau,x,k\right). \tag{2.9}$$

Here, we wrote all the arguments of the function $u^{\mathrm{BS}}\left(\sigma_0^{(\bar{\mathbf{z}})};\tau,x,k\right)$ to show that it does not depend on y and z.

Note that $a_{11}(\mathbf{z}) = -a_1(\mathbf{z})$. It follows that

$$\begin{aligned}\mathscr{G}_{i_h}^{(\bar{\mathbf{z}})}(t,s_h,\mathbf{z}) u^{\mathrm{BS}}\left(\sigma_0^{(\bar{\mathbf{z}})};\tau,x,k\right) &= a_{11,i_h}(\mathbf{z} - \bar{\mathbf{z}} + \mathbf{m}^{(\bar{\mathbf{z}})}(t,s_h) + C^{(\bar{\mathbf{z}})}(t,s_h)\nabla_{\mathbf{z}}) \\ &\quad \times \left(\frac{\partial^2}{\partial x^2} - \frac{\partial}{\partial x}\right) u^{\mathrm{BS}}\left(\sigma_0^{(\bar{\mathbf{z}})};\tau,x,k\right),\end{aligned}$$

and equation [2.9] can be written as

$$u_n^{(\bar{\mathbf{z}})}(t,\mathbf{z};T,k) = \tilde{\mathscr{L}}_n^{(\bar{\mathbf{z}})}(t,T,\mathbf{z}) \left(\frac{\partial^2}{\partial x^2} - \frac{\partial}{\partial x}\right) u^{\mathrm{BS}}\left(\sigma_0^{(\bar{\mathbf{z}})};\tau,x,k\right),$$

where the operator $\tilde{\mathscr{L}}_n^{(\bar{\mathbf{z}})}(t,T,\mathbf{z})$ is given by (Lorig *et al*. 2017, Equation 3.14) as

$$\begin{aligned}\tilde{\mathscr{L}}_n^{(\bar{\mathbf{z}})}(t,T,\mathbf{z}) = \sum_{h=1}^{n} \int_t^T \int_{s_1}^T \cdots \int_{s_{h-1}}^T \sum_{\mathbf{i} \in I_{n,h}} \mathscr{G}_{i_1}^{(\bar{\mathbf{z}})}(t,s_1,\mathbf{z}) \cdots \mathscr{G}_{i_{h-1}}^{(\bar{\mathbf{z}})}(t,s_{h-1},\mathbf{z}) \\ \times a_{11,i_h}(\mathbf{z} - \bar{\mathbf{z}} + \mathbf{m}^{(\bar{\mathbf{z}})}(t,s_h) + C^{(\bar{\mathbf{z}})}(t,s_h)\nabla_{\mathbf{z}})\, \mathrm{d}s_h \cdots \mathrm{d}s_1.\end{aligned} \tag{2.10}$$

It is well-known that

$$\frac{\partial}{\partial \sigma} u^{\mathrm{BS}}\left(\sigma_0^{(\bar{\mathbf{z}})};\tau,x,k\right) = \sigma_0^{(\bar{\mathbf{z}})}\tau \left(\frac{\partial^2}{\partial x^2} - \frac{\partial}{\partial x}\right) u^{\mathrm{BS}}\left(\sigma_0^{(\bar{\mathbf{z}})};\tau,x,k\right).$$

The first term on the right-hand side of equation [2.6] takes the form of (Lorig *et al*. 2017, Equation 3.13)

$$\begin{aligned}u_n^{(\bar{\mathbf{z}})}\left(\frac{\partial}{\partial \sigma} u^{\mathrm{BS}}\left(\sigma_0^{(\bar{\mathbf{z}})}\right)\right)^{-1} &= \tilde{\mathscr{L}}_n^{(\bar{\mathbf{z}})}(t,T,\mathbf{z}) \left(\frac{\partial^2}{\partial x^2} - \frac{\partial}{\partial x}\right) u^{\mathrm{BS}}\left(\sigma_0^{(\bar{\mathbf{z}})};\tau,x,k\right) \\ &\quad \times \left[\sigma_0^{(\bar{\mathbf{z}})}\tau \left(\frac{\partial^2}{\partial x^2} - \frac{\partial}{\partial x}\right) u^{\mathrm{BS}}\left(\sigma_0^{(\bar{\mathbf{z}})};\tau,x,k\right)\right]^{-1}.\end{aligned}$$

It follows that there exist functions $\chi_m^{(n)}(\bar{\mathbf{z}};t,T)$ such that

$$u_n^{(\bar{\mathbf{z}})}\left(\frac{\partial}{\partial\sigma}u^{\mathrm{BS}}\left(\sigma_0^{(\bar{\mathbf{z}})}\right)\right)^{-1} = \sum_m \chi_m^{(n)}(\bar{\mathbf{z}};t,T)\frac{\partial^m}{\partial x^m}\left(\frac{\partial^2}{\partial x^2}-\frac{\partial}{\partial x}\right)u^{\mathrm{BS}}\left(\sigma_0^{(\bar{\mathbf{z}})};\tau,x,k\right) \times\left[\left(\frac{\partial^2}{\partial x^2}-\frac{\partial}{\partial x}\right)u^{\mathrm{BS}}\left(\sigma_0^{(\bar{\mathbf{z}})};\tau,x,k\right)\right]^{-1}.$$

(see Lorig *et al.* (2017), Equation 3.15). This is because the function $u^{\mathrm{BS}}\left(\sigma_0^{(\bar{\mathbf{z}})};\tau,x,k\right)$ does not depend on y and z.

From Lorig *et al.* (2017), Lemma 3.4, we have

$$\frac{\partial^m}{\partial x^m}\left(\frac{\partial^2}{\partial x^2}-\frac{\partial}{\partial x}\right)u^{\mathrm{BS}}(\sigma_0)\left[\left(\frac{\partial^2}{\partial x^2}-\frac{\partial}{\partial x}\right)u^{\mathrm{BS}}(\sigma_0)\right]^{-1} = \left(-\frac{1}{\sigma_0\sqrt{2\tau}}\right)^m H_m(\zeta), \quad [2.11]$$

where

$$\zeta = \frac{x-k-\sigma_0^2\tau/2}{\sigma_0\sqrt{2\tau}},$$

and where

$$H_m(\zeta) = (-1)^m\exp(\zeta^2)\frac{\partial^m}{\partial\zeta^m}\exp(-\zeta^2)$$

is the mth Hermite polynomial.

We must still calculate the expression in the third line of equation [2.6] for $h \geq 2$ (it is equal to 1 when $h=1$). From Lorig *et al.* (2017), Proposition 3.5, we have

$$\frac{\partial^h}{\partial\sigma^h}u^{\mathrm{BS}}(\sigma_0)\left(\frac{\partial}{\partial\sigma}u^{\mathrm{BS}}(\sigma_0)\right)^{-1} = \sum_{q=0}^{\lfloor h/2\rfloor} c_{h,h-2q}\sigma_0^{h-2q-1}\tau^{h-q-1}\sum_{p=0}^{h-q-1}\binom{h-q-1}{p} \times\left(\frac{1}{\sigma_0\sqrt{2\tau}}\right)^{p+h-q-1}H_{p+h-q-1}(\zeta), \quad [2.12]$$

where the coefficients $c_{h,h-2q}$ are defined recursively by

$$c_{h,h}=1, \qquad c_{h,h-2q} = (h-2q+1)c_{h-1,h-2q+1}+c_{h-1,h-2q-1}.$$

Using equations [2.7] and [2.10], we explicitly calculate:

$$a_{11,1}(\mathbf{z}) = \frac{1}{2}(y - v_0)$$

and

$$\begin{aligned}\mathscr{L}_1^{(\bar{\mathbf{z}})}(t,T,\mathbf{z}) &= \frac{1}{2}(T-t)(y-v_0) + \frac{1}{4}(T-t)^2\kappa_1(v_0'-v_0)\\ &\quad + \frac{1}{4}(T-t)^2\rho_{12}\xi_1 v_0^{\alpha_1+1/2}\frac{\partial}{\partial x} + \cdots,\end{aligned}$$

where the dots denote the terms containing $\frac{\partial}{\partial y}$ and $\frac{\partial}{\partial z}$. The functions $\chi_m^{(1)}(\bar{\mathbf{z}};t,T)$ take the form

$$\begin{aligned}\chi_0^{(1)}(\bar{\mathbf{z}};t,T) &= \frac{y-v_0}{2\sqrt{v_0}} + \frac{(T-t)\kappa_1(v_0'-v_0)}{4\sqrt{v_0}},\\ \chi_1^{(1)}(\bar{\mathbf{z}};t,T) &= \frac{(T-t)\rho_{12}\xi_1 v_0^{\alpha_1+1/2}}{4\sqrt{v_0}}.\end{aligned}$$

Equation [2.6] gives

$$\begin{aligned}\sigma_1^{(\bar{\mathbf{z}})}(t,\mathbf{z};T,k) &= \frac{y-v_0}{2\sqrt{v_0}} + \frac{(T-t)\kappa_1(v_0'-v_0)}{4\sqrt{v_0}}\\ &\quad - \frac{\rho_{12}\xi_1 v_0^{\alpha_1+1/2}(x-k-\sigma_0^2(T-t)/2)}{4v_0^{3/2}}.\end{aligned}$$

Then,

$$\sigma_1^{(\mathbf{z})}(t,\mathbf{z};T,k) = \frac{(T-t)\kappa_1(z-y)}{4\sqrt{v_0}} - \frac{\rho_{12}\xi_1 y^{\alpha_1+1/2}(x-k-\sigma_0^2(T-t)/2)}{4v_0^{3/2}}$$

and

$$\overline{\sigma}_1(t,\mathbf{z};T,k) = \sqrt{v_0} + \frac{(T-t)\kappa_1(z-y)}{4\sqrt{v_0}} - \frac{\rho_{12}\xi_1 y^{\alpha_1+1/2}(x-k-\sigma_0^2(T-t)/2)}{4v_0^{3/2}}.$$

As $T \to t$ and $k \to x$, the second and third terms disappear. Calculating the derivative with respect to k, we obtain

$$\frac{\partial}{\partial k}\overline{\sigma}_1(t,\mathbf{z};T,k) = \frac{\rho_{12}\xi_1 v_0^{\alpha_1+1/2}}{4v_0^{3/2}},$$

and Theorem 2.2 follows.

PROOF OF THEOREM 2.3.–

Equation [2.6] takes the form

$$\begin{aligned}\sigma_2^{(\bar{\mathbf{z}})}(t,x,y,z;T,k) &= u_2^{(\bar{\mathbf{z}})}\left(\frac{\partial}{\partial\sigma}u^{\mathrm{BS}}\left(\sigma_0^{(\bar{\mathbf{z}})}\right)\right)^{-1} \\ &\quad -\frac{1}{2}(\sigma_1^{(\bar{\mathbf{z}})})^2\frac{\partial^2}{\partial\sigma^2}u^{\mathrm{BS}}\left(\sigma_0^{(\bar{\mathbf{z}})}\right)\left(\frac{\partial}{\partial\sigma}u^{\mathrm{BS}}\left(\sigma_0^{(\bar{\mathbf{z}})}\right)\right)^{-1}.\end{aligned} \quad [2.13]$$

The sets $I_{2,h}$ are $I_{2,1}=\{(2)\}$, $I_{2,2}=\{(1,1)\}$. We have $a_{11,2}(x,y,z)=0$. It follows that equation [2.10] with $n=2$ includes only summation over the set $I_{2,2}$ and takes the form

$$\begin{aligned}\tilde{\mathscr{L}}_2^{(\bar{\mathbf{z}})}(t,T,\mathbf{z}) &= \int_t^T\int_{t_1}^T \mathscr{G}_1^{(\bar{\mathbf{z}})}(t,t_1,\mathbf{z}) \\ &\quad\times a_{11,1}(\mathbf{z}-\bar{\mathbf{z}}+\mathbf{m}^{(\bar{\mathbf{z}})}(t,t_2)+C^{(\bar{\mathbf{z}})}(t,t_2)\nabla_{\mathbf{z}})\,\mathrm{d}t_2\,\mathrm{d}t_1.\end{aligned}$$

While calculating the operator $\mathscr{G}_1^{(\bar{\mathbf{z}})}(t,t_1,\mathbf{z})$ using equation [2.8], we need to calculate only the coefficients of the three partial derivatives with respect to the variable x. We obtain

$$\begin{aligned}\mathscr{G}_1^{(\bar{\mathbf{z}})}(t,t_1,\mathbf{z}) &= \frac{1}{2}(t_1-t)\rho_{12}\xi_1 v_0^{\alpha_1+1/2}\frac{\partial^3}{\partial x^3} \\ &\quad+\left[\frac{1}{2}(y-v_0)+\frac{1}{2}(t_1-t)\kappa_1(v_0'-v_0)-\frac{1}{4}(t_1-t)\rho_{12}\xi_1 v_0^{\alpha_1+1/2}\right]\frac{\partial^2}{\partial x^2} \\ &\quad-\frac{1}{2}[(y-v_0)+(t_1-t)\kappa_1(v_0'-v_0)]\frac{\partial}{\partial x}+\cdots.\end{aligned}$$

The following integrals are important for calculations:

$$\begin{aligned}\int_t^T\int_{t_1}^T (t_1-t)(t_2-t)\,\mathrm{d}t_2\,\mathrm{d}t_1 &= \frac{1}{8}(T-t)^4, \\ \int_t^T\int_{t_1}^T (t_1-t)\,\mathrm{d}t_2\,\mathrm{d}t_1 &= \frac{1}{6}(T-t)^3, \\ \int_t^T\int_{t_1}^T (t_2-t)\,\mathrm{d}t_2\,\mathrm{d}t_1 &= \frac{1}{3}(T-t)^3, \\ \int_t^T\int_{t_1}^T \mathrm{d}t_2\,\mathrm{d}t_1 &= \frac{1}{2}(T-t)^2.\end{aligned}$$

The operator $\mathscr{L}_2^{(\bar{\mathbf{z}})}(t,T,\mathbf{z})$ takes the form

$$\begin{aligned}\mathscr{L}_2^{(\bar{\mathbf{z}})}(t,T,\mathbf{z}) &= \frac{1}{32}(T-t)^4\rho_{12}^2\xi_1^2 v_0^{2\alpha_1+1}\frac{\partial^4}{\partial x^4}\\ &+\frac{1}{32}[2(T-t)^4\rho_{12}\xi_1 v_0^{\alpha_1+1/2}\kappa_1(v_0'-v_0)\\ &+4(T-t)^3\rho_{12}\xi_1 v_0^{\alpha_1+1/2}(y-v_0)-(T-t)^4\rho_{12}^2\xi_1^2 v_0^{2\alpha_1+1}]\frac{\partial^3}{\partial x^3}\\ &+\frac{1}{32}[(T-t)^4\kappa_1^2(v_0'-v_0)^2-2(T-t)^4\rho_{12}\xi_1 v_0^{\alpha_1+1/2}\kappa_1(v_0'-v_0)\\ &+4(T-t)^3\kappa_1(v_0'-v_0)(y-v_0)-4(T-t)^3\rho_{12}\xi_1 v_0^{\alpha_1+1/2}(y-v_0)\\ &+4(T-t)^2(y-v_0)^2]\frac{\partial^2}{\partial x^2}\\ &-\frac{1}{32}[(T-t)^4\kappa_1^2(v_0'-v_0)^2+4(T-t)^3\kappa_1(v_0'-v_0)(y-v_0)\\ &+4(T-t)^2(y-v_0)^2]\frac{\partial}{\partial x}+\dots.\end{aligned}$$

Calculation of the first term on the right-hand side of equation [2.13] using equation [2.11] may be left to the reader.

Next, we calculate the left-hand side of equation [2.12] for $h=2$. Using the Hermite polynomials $H_0(\zeta)=1$, $H_1(\zeta)=2\zeta$ and $H_2(\zeta)=4\zeta^2-2$, we obtain

$$\frac{\partial^2}{\partial\sigma^2}u^{\mathrm{BS}}(\sigma_0)\left(\frac{\partial}{\partial\sigma}u^{\mathrm{BS}}(\sigma_0)\right)^{-1}=\sqrt{2(T-t)}\zeta+2\sigma_0^{-1}\zeta^2.$$

Combining everything together, we obtain the formula for $\overline{\sigma}_2(t,x,y,z;T,k)$

$$\begin{aligned}\overline{\sigma}_2(t,x,y,z;T,k) &= \sqrt{v_0}+\frac{\kappa_1(v_0'-v_0)}{4\sqrt{v_0}}(T-t)-\frac{1}{4}\rho_{12}\xi_1 v_0^{\alpha_1-1}(x_0-k)\\ &+\frac{1}{16}\rho_{12}\xi_1 v_0^{\alpha_1}(T-t)-\frac{3}{32}\rho_{12}^2\xi_1^2 v_0^{2\alpha_1-2}(x_0-k)^2\\ &+\frac{3}{32}\rho_{12}^2\xi_1^2 v_0^{2\alpha_1-3/2}(T-t)+\cdots,\end{aligned}\qquad [2.14]$$

where the ellipsis denotes the terms satisfying the following condition: the limits of the term, its first partial derivative with respect to T and its first two partial derivatives with respect to k as (T,k) approaches (t,x) within $\mathscr{P}_\lambda$, are all equal to 0.

On the right-hand side of equation [2.14], the first term, the partial derivatives with respect to T of the second, fourth and sixth terms, the first partial derivative with respect to k of the third term, and the second partial derivative with respect to k of the fifth term give nonzero contributions to the right-hand side of the asymptotic expansion [2.4].

2.4. References

Gatheral, J. (2008). Consistent modelling of SPX and VIX options. *The Fifth World Congress of the Bachelier Finance Society*, London.

Latané, H.A. and Rendleman Jr., R.J. (1976). Standard deviations of stock price ratios implied in option prices. *J. Finance*, 31(2), 369–381.

Lorig, M., Pagliarani, S., Pascucci, A. (2017). Explicit implied volatilities for multifactor local-stochastic volatility models. *Math. Finance*, 27(3), 926–960.

Orlando, G. and Taglialatela, G. (2017). A review on implied volatility calculation. *J. Comput. Appl. Math.*, 320, 202–220.

Pagliarani, S. and Pascucci, A. (2012). Analytical approximation of the transition density in a local volatility model. *Cent. Eur. J. Math.*, 10(1), 250–270.

Pagliarani, S. and Pascucci, A. (2017). The exact Taylor formula of the implied volatility. *Finance Stoch.*, 21(3), 661–718.

3

New Dividend Strategies

We will consider two insurance models with dividend payments. The first one is the Cramér–Lundberg model with exponentially distributed claims. We will study a barrier dividend strategy with the Parisian implementation delay. This means that if the company surplus stays a prescribed time interval h above the barrier, the overshoot is immediately paid as dividend. The expected discounted dividends paid before the Parisian ruin are chosen as the objective function. Numerical results are provided. The second model is the dual Cramér–Lundberg model with exponentially distributed gains. Instead of barrier strategy, we deal with a threshold one, meaning that dividends are paid with a constant rate as long as the surplus stays above the threshold.

3.1. Introduction

Insurance has a long and interesting history. Methods for transferring or distributing risks were practiced by Chinese and Babylonian traders more than 3000 years BCE. Let us mention, for example, the Code of Hammurabi c. 1750 BCE, which dealt with maritime risks. Mutual societies, run by their members with no external shareholders to pay, were the first to appear. The next step is joint stock companies. This explains the two-fold nature of modern insurance companies. The primary task of a company is indemnification of policyholder claims. The secondary, but very important, task is dividend payments to shareholders.

It is well known that insurance models are of the input–output type. We have to specify the premiums inflow $P(t)$ and claim payments to customers (outflow) $S(t)$, as well as the planning horizon $T \leq \infty$. Thus, the company surplus (capital or reserve) $X(t)$ at time $t \leq T$ has the form $X(t) = x + P(t) - S(t)$. Here, x is the initial surplus. To accomplish the optimization of the insurance company performance, we

Chapter written by Ekaterina BULINSKAYA.

need to introduce the set of feasible controls and an objective function. It is possible to use different objective functions (criteria, targets or risk measures) in order to evaluate an insurance company's performance. The most popular one in non-life insurance (since 1903) was the company ruin probability for the classical (collective risk) Cramér–Lundberg model. In other words, the main goal was to achieve the high reliability of the company. In practice, it turned out that the negative surplus level may not always lead to bankruptcy, since the company may use, for example, a bank loan to avoid insolvency. Therefore, there were defined and studied "Absolute ruins", "Parisian ruins", as well as, "Omega models", in the framework of reliability approach (for the definitions, see Bulinskaya (2017) and the references therein). New problems have arisen in actuarial sciences during the last 20 years. This period is characterized by the interplay of insurance and finance, the unification of reliability and the cost approaches (see, for example, Bulinskaya (2017)), as well as the consideration of complex systems. Sophisticated mathematical tools are used for the analysis and optimization of insurance systems including dividends, reinsurance and investment.

A dividend is a payment made by a corporation to its shareholders, usually as a distribution of profits. It was Bruno de Finetti who introduced the dividends study in actuarial mathematics in 1957. In his seminal paper, de Finetti (1957) argued that under the net profit condition, the insurance company surplus could become infinite as time grows. As this is unrealistic, it is necessary to choose a dividend strategy. That was the beginning of cost approaches in actuarial mathematics.

A large number of papers have been devoted to dividends payment (see, for example, Avanzi (2009), Albrecher and Thonhauser (2009) for the survey of the results published before 2009). Expected discounted dividends paid before ruin are usually taken as the objective function (see the classical textbooks by Bühlman (1970) and Gerber (1979), as well as the paper by Sethi *et al.* (1991)). The barrier strategy is the most popular, although in Azcue and Muler (2005), it was established that the optimal dividend strategy is not always the barrier strategy. Many ramifications of this strategy were proposed (see, for example, Drekic *et al.* (2018) and the references therein).

Below, we investigate two models with dividends. In section 3.2, we treat the classical Cramér–Lundberg insurance model with exponential claims and the barrier dividend strategy generalizing those introduced in Dassios and Wu (2009) and Bulinskaya and Shigida (2018). The attention is focused on the calculation of the objective function and simulation. In section 3.3, we study the dual Cramér–Lundberg model with exponential gains and the threshold dividend strategy. Integro-differential equations for the expected discounted dividends are established. Using the explicit form of the objective function, we obtained the optimal threshold. Section 3.4 contains the conclusion and further research directions.

3.2. Model 1

We consider the standard Cramér–Lundberg model

$$X_t = X_0 + ct - \sum_{i=1}^{N_t} C_i, \quad t \geq 0, \qquad [3.1]$$

where X_t is the insurance company surplus at time t, c is the rate of premium accumulation, $\{N_t\}_{t\geq 0}$ is the Poisson process with intensity λ and the claim amounts $\{C_i\}_{i=1}^{\infty}$ are independent identically distributed random variables, independent of the Poisson process. Additionally, assume that the claim distribution is exponential with parameter β.

We will use the fact that X_t has independent increments and is translation invariant; therefore, the strong Markov property is applicable. Also, $X_t - \mathsf{E}X_t$ is a martingale (because it is a process with independent increments whose mean value is constant), and the optimal stopping theorem is applicable as well.

LEMMA 3.1.– *If the condition*

$$c \leq \frac{\lambda}{\beta}$$

holds, then for any initial capital $x > 0$*, ruin happens with probability 1.*

The proof can be found in Mikosch (2006).

Therefore, further on, we suppose the net profit condition $c > \frac{\lambda}{\beta}$ to be fulfilled. We also need some results that were established in Bulinskaya and Shigida (2019).

Let d and h take non-negative real values and $l_i \in \mathbf{R}^1$, $i = 1, 2$. Everywhere, l_1 is assumed to be less than l_2. Put

$$\tau_{x,l_2} = \inf\{t \geq 0 \mid X_0 = x, X_t = l_2\},$$

$$T_{x,l_1,d} = \inf\{t \geq 0 \mid X_0 = x, X_t \text{ has been} < l_1 \text{ for at least } d\},$$

$$F_{x,l_2,h} = \inf\{t \geq 0 \mid X_0 = x, X_t \text{ has been} \geq l_2 \text{ for at least } h\}.$$

Let $r > 0$ be the force of interest, and v_r^+ and v_r^- be the positive and negative roots of the equation, respectively

$$-r + cv_r + \lambda\left(\frac{\beta}{\beta + v_r} - 1\right) = 0.$$

Let U_i be the ith excursion of the process X_t above l_1 and V_i be the ith excursion below l_1. All of these random variables are independent (by the strong Markov property). If $x \geq l_1$, U_1 has a distribution different from that of other $\{U_i\}_{i=2}^{\infty}$, which are identically distributed, so are $\{V_i\}_{i=1}^{\infty}$. If $x < l_1$, V_1 has a distribution different from that of other $\{V_i\}_{i=2}^{\infty}$, which are identically distributed, so are $\{U_i\}_{i=1}^{\infty}$. In any case, let p_1 be the density of U_2 and p_2 be the density of V_2 (p_1 and p_2 do not depend on l_1). If $x \geq l_1$, let $g_1^{x-l_1}$ be the density of U_1 (this way, g_1^x does not depend on l_1), otherwise let $g_2^{x-l_1}$ be the density of V_1 (g_2^x also does not depend on l_1).

Now, we can formulate the needed results.

LEMMA 3.2.– *For* $l_1 \leq x < l_2$

$$\mathsf{E}e^{-r\tau_{x,l_2}}\mathbf{1}_{\{\tau_{x,l_2}<T_{x,l_1,d}\}} = \frac{(\beta+v_r^+)e^{v_r^+(x-l_1)}-(\beta+v_r^-)e^{v_r^-(x-l_1)}}{(\beta+v_r^+)e^{v_r^+(l_2-l_1)}-(\beta+v_r^-)e^{v_r^-(l_2-l_1)}}$$

$$+\frac{v_r^+-v_r^-}{(\beta+v_r^+)e^{v_r^+(l_2-l_1)}-(\beta+v_r^-)e^{v_r^-(l_2-l_1)}}$$

$$\times\frac{\left(e^{v_r^+(x-l_2)}-e^{v_r^-(x-l_2)}\right)\int_0^d e^{-rs}p_2(s)\,ds}{\left(\frac{\beta e^{v_r^+(l_1-l_2)}}{\beta+v_r^+}-\frac{\beta e^{v_r^-(l_1-l_2)}}{\beta+v_r^-}\right)-\left(e^{v_r^+(l_1-l_2)}-e^{v_r^-(l_1-l_2)}\right)\int_0^d e^{-rs}p_2(s)\,ds}.$$

For $x < l_1$

$$\mathsf{E}e^{-r\tau_{x,l_2}}\mathbf{1}_{\{\tau_{x,l_2}<T_{x,l_1,d}\}} = \frac{v_r^+-v_r^-}{(\beta+v_r^+)e^{v_r^+(l_2-l_1)}-(\beta+v_r^-)e^{v_r^-(l_2-l_1)}}$$

$$\times\left(\frac{\beta e^{v_r^+(l_1-l_2)}}{\beta+v_r^+}-\frac{\beta e^{v_r^-(l_1-l_2)}}{\beta+v_r^-}\right)$$

$$\times\frac{\int_0^d e^{-rs}g_2^{x-l_1}(s)\,ds}{\left(\frac{\beta e^{v_r^+(l_1-l_2)}}{\beta+v_r^+}-\frac{\beta e^{v_r^-(l_1-l_2)}}{\beta+v_r^-}\right)-\left(e^{v_r^+(l_1-l_2)}-e^{v_r^-(l_1-l_2)}\right)\int_0^d e^{-rs}p_2(s)\,ds}.$$

Let us now introduce another level $l_0 < l_1$.

THEOREM 3.1.– *The following relation is valid:*

$$\mathsf{E}e^{-rF_{l_1,l_1,h}}\mathbf{1}_{\{F_{l_1,l_1,h}<T_{l_1,l_0,d}\}} = \frac{e^{-rh}\bar{P}_1(h)}{1-\mathbf{A}\times\int_0^h e^{-rs}p_1(s)\,ds},$$

where

$$\mathbf{A} = \frac{\beta e^{v_r^+(l_1-l_0)} - \beta e^{v_r^-(l_1-l_0)}}{(\beta+v_r^+)e^{v_r^+(l_1-l_0)} - (\beta+v_r^-)e^{v_r^-(l_1-l_0)}}$$
$$+ \frac{v_r^+ - v_r^-}{(\beta+v_r^+)e^{v_r^+(l_1-l_0)} - (\beta+v_r^-)e^{v_r^-(l_1-l_0)}} \times \left(\frac{\beta}{\beta+v_r^+} - \frac{\beta}{\beta+v_r^-}\right)$$
$$\times \frac{\int_0^d e^{-rs}p_2(s)\,ds}{\left(\frac{\beta e^{v_r^+(l_0-l_1)}}{\beta+v_r^+} - \frac{\beta e^{v_r^-(l_0-l_1)}}{\beta+v_r^-}\right) - \left(e^{v_r^+(l_0-l_1)} - e^{v_r^-(l_0-l_1)}\right)\int_0^d e^{-rs}p_2(s)\,ds}.$$

REMARK.– This theorem generalizes Theorem 3.1 in Bulinskaya and Shigida (2018).

THEOREM 3.2.– *The function* $\mathsf{E}e^{-rF_{l_1,l_1,h}}X_{F_{l_1,l_1,h}}\mathbf{1}_{\{F_{l_1,l_1,h}<T_{l_1,l_0,d}\}}$ *is given by*

$$\mathsf{E}e^{-rF_{l_1,l_1,h}}\mathbf{1}_{\{F_{l_1,l_1,h}<T_{l_1,l_0,d}\}}\times$$
$$\left(\frac{\frac{1}{\beta}+\left(c-\frac{\lambda}{\beta}\right)\int_0^h sp_1(s)\,ds}{\bar{P}_1(h)} + l_1 + ch - \frac{\lambda h+1}{\beta}\right).$$

The expectation of total dividend payments until the Parisian ruin, under barrier strategy, with the Parisian implementation delay is our next goal.

Thus, we consider a company whose capital at time t (before dividends payment) is X_t. In order to formulate a new dividend strategy with payment delay generalizing that introduced in Dassios and Wu (2009), we need some notations. The initial surplus is $X_0 = x$, where $0 \le x < b$. We set $g_{b,t}^X = \sup\{s \le t \mid X_s \le b\}$. Let

$$\tau_0^X = \inf\{t \ge 0 \mid X_t = b\}$$

be the first time the process $\{X_t\}$ hits the barrier b and let

$$\tau_i^X = \inf\left\{t \ge \tau_{i-1}^X \mid \mathbf{1}_{\left\{X_t > X_{\tau_{i-1}^X}\right\}}\left(t - g_{X_{\tau_{i-1}^X},t}^X\right) \ge h\right\}$$

be the first time after τ_{i-1}^X when the length of the excursion above $X_{\tau_{i-1}^X}$ reaches h. We assume that dividends are paid only if the surplus stayed above the barrier b during the time interval of length h. Then, the excess is immediately paid out and the surplus starts from level b.

The modified surplus process (taking into account dividend payments)

$$Y_t = X_t \mathbf{1}_{\{0 \le t < \tau_0^X\}} + \sum_{i=0}^{\infty} \left(X_t - X_{\tau_i^X} + b \right) \mathbf{1}_{\{\tau_i^X \le t < \tau_{i+1}^X\}}. \quad [3.2]$$

Define the time of Parisian ruin by

$$T_d = \inf \left\{ t > 0 \mid \mathbf{1}_{\{Y_t < 0\}} \left(t - d_t^Y \right) \ge d \right\},$$

where $d_t^Y = \sup \left\{ s < t \mid Y_s \ge 0 \right\}$. The present value of the total dividend payments before the Parisian ruin of the modified process is given by

$$V(x, b) = \sum_{i=1}^{\infty} e^{-r\tau_i^X} \left(X_{\tau_i^X} - X_{\tau_{i-1}^X} \right) \mathbf{1}_{\{\tau_i^X \le T_d\}}. \quad [3.3]$$

Obviously, $\mathsf{E}V(x, b)$ depends not only on the initial surplus x and dividends barrier b. However, other parameters (d, h, r, λ, β) are omitted in order to simplify the notation.

Also, denote by $\mathrm{NR}\left[t_1, t_2\right)$ the event that there is no moment $t_1 \le t < t_2$, when the surplus has stayed below zero for at least d. Then, we can rewrite [3.3] as

$$V(x, b) = \sum_{i=1}^{\infty} e^{-r\tau_i^X} \left(X_{\tau_i^X} - X_{\tau_{i-1}^X} \right) \mathbf{1}_{\mathrm{NR}\left[0, \tau_i^X\right)}.$$

THEOREM 3.3.– *The following relation:*

$$\mathsf{E}V(x, b) = \mathsf{E}e^{-r\tau_0^X} \mathbf{1}_{NR\left[0, \tau_0^X\right)} \frac{\mathsf{E}e^{-r\left(\tau_1^X - \tau_0^X\right)} \left(X_{\tau_1^X} - X_{\tau_0^X} \right) \mathbf{1}_{NR\left[\tau_0^X, \tau_1^X\right)}}{1 - \mathsf{E}e^{-r\left(\tau_1^X - \tau_0^X\right)} \mathbf{1}_{NR\left[\tau_0^X, \tau_1^X\right)}} \quad [3.4]$$

is true.

THE SKETCH OF PROOF.– By the strong Markov property, we can write $\mathsf{E}V(x, b)$ as

$$\begin{aligned} &\mathsf{E}e^{-r\tau_0^X} \mathbf{1}_{\mathrm{NR}\left[0, \tau_0^X\right)} \mathsf{E} \sum_{i=1}^{\infty} e^{-r\left(\tau_i^X - \tau_0^X\right)} \left(X_{\tau_i^X} - X_{\tau_{i-1}^X} \right) \mathbf{1}_{\mathrm{NR}\left[\tau_0^X, \tau_i^X\right)} \\ &\qquad = \mathsf{E}e^{-r\tau_0^X} \mathbf{1}_{\mathrm{NR}\left[0, \tau_0^X\right)} \mathsf{E}V(b). \end{aligned} \quad [3.5]$$

For now, let us concentrate on finding $\mathsf{E}V(b)$, where $V(b) = V(b, b)$

$$\begin{aligned}\mathsf{E}V(b) &= \mathsf{E}e^{-r(\tau_1^X - \tau_0^X)}\left(X_{\tau_1^X} - X_{\tau_0^X}\right)\mathbf{1}_{\mathrm{NR}[\tau_0^X, \tau_1^X)}\\ &+ \mathsf{E}e^{-r(\tau_1^X - \tau_0^X)}\mathbf{1}_{\mathrm{NR}[\tau_0^X, \tau_1^X)}\mathsf{E}\sum_{i=2}^{\infty} e^{-r(\tau_i^X - \tau_0^X)}\left(X_{\tau_i^X} - X_{\tau_{i-1}^X}\right)\mathbf{1}_{\mathrm{NR}[\tau_0^X, \tau_i^X)}\\ &= \mathsf{E}e^{-r(\tau_1^X - \tau_0^X)}\left(X_{\tau_1^X} - X_{\tau_0^X}\right)\mathbf{1}_{\mathrm{NR}[\tau_0^X, \tau_1^X)}\\ &\qquad + \mathsf{E}e^{-r(\tau_1^X - \tau_0^X)}\mathbf{1}_{\mathrm{NR}[\tau_0^X, \tau_1^X)}\mathsf{E}V(b).\end{aligned}$$

Solving this simple linear equation with respect to $\mathsf{E}V(b)$, we get

$$\mathsf{E}V(b) = \frac{\mathsf{E}e^{-r(\tau_1^X - \tau_0^X)}\left(X_{\tau_1^X} - X_{\tau_0^X}\right)\mathbf{1}_{\mathrm{NR}[\tau_0^X, \tau_1^X)}}{1 - \mathsf{E}e^{-r(\tau_1^X - \tau_0^X)}\mathbf{1}_{\mathrm{NR}[\tau_0^X, \tau_1^X)}}.$$

Recalling [3.5], we establish the desired result, thus ending the proof. □

The explicit expression of the function $\mathsf{E}V(x, b)$ can be obtained as follows. Three terms in [3.4] are given by the first expression in Lemma 3.2 with $l_1 = 0$, $l_2 = b$ and in Theorems 3.2 and 3.3 with $l_0 = 0$, $l_1 = b$.

Our task is to find the optimal barrier b^* maximizing the expectation $\mathsf{E}V(x, b)$. Analysis of $\mathsf{E}V(x, b)$ also provides the following result.

COROLLARY 3.1.– *For any* $d > 0$, $\mathsf{E}V(x, b) > \mathsf{E}V(x, b)|_{d=0}$.

Thus, it is possible to establish that the expected present value of dividends until the Parisian ruin ($d > 0$) is greater than that until the classical ruin ($d = 0$).

The explicit expression [3.4] of the function $\mathsf{E}V(x, b)$ seems very complicated for analytical investigation. Therefore, the numerical results were obtained first. An analysis of the model under consideration was conducted using the Python programming language. In Figure 3.1, we provide six graphs of the expected discounted total dividend payment as a function of b (for $c = 10.0$, $\lambda = 5.4$, $\beta = 1.0$, $d = 0.1$, $h = 0.3$, $r = 0.2$ and $x = 0, 1, 2, 3, 4, 5$).

It can be seen from those graphs that the optimal barrier b^* maximizing the expectation (see the vertical yellow line) does not depend on x. The function was analyzed in Python using the `scipy` library, and the graphs were obtained with the `matplotlib` library.

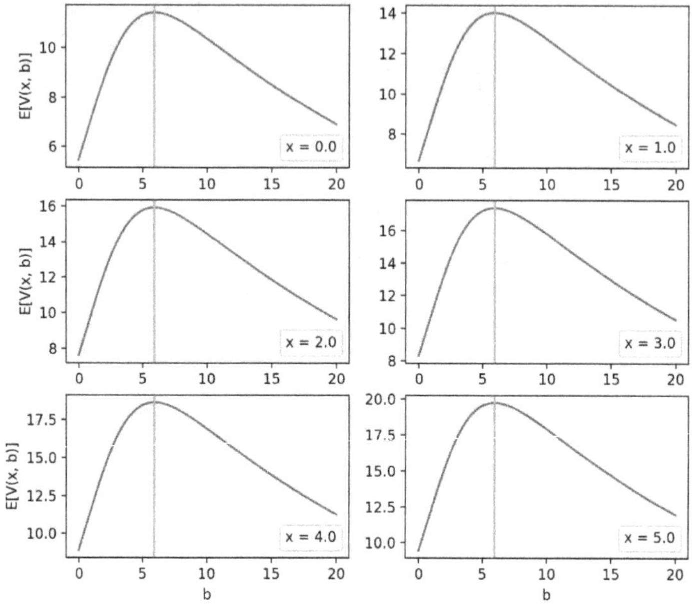

Figure 3.1. *Form of* $\mathsf{E}V(x,b)$ *as function of* b *for fixed* x. *For a color version of this figure, see www.iste.co.uk/dimotikalis/analysis2.zip*

Using the first expression in Lemma 3.2, it can be shown that $\mathsf{E}e^{-r\tau_0^X}\mathbf{1}_{\text{NR}[0,\tau_0^X)}$ has the form $C(x)f(b)$, where the function $f(b)$ does not depend on x and the factor $C(x)$ does not depend on b.

Thus, by taking the derivative of the function $\mathsf{E}V(x,b)$ with respect to b and by finding the point b^* where it equals zero, we can find the global maximum of this function.

Due to [3.4], this means that the partial derivative with respect to b of the expected total dividend payment has such a form as well

$$\frac{\partial}{\partial b}\mathsf{E}V(x,b) = C(x)\tau(b),$$

where $\tau(b)$ does not depend on x. This means that the optimal barrier b^*, satisfying

$$\tau(b^*) = 0,$$

does not depend on x either.

If $b^* > 0$ (which does not follow from the net profit condition), for all $0 \leq x < b^*$, it is the optimal barrier (the barrier that maximizes the expected total dividend payment).

Also, a simulation of the process Y_t itself was carried out. We generated a large sample of independent exponential random variables with parameter λ, which are treated as intervals between claims, and of independent exponential random variables with parameter β, which represent claim amounts. Random samples were generated using the standard module `random` in Python.

The formulas [3.2] and [3.1] translated into code were applied directly, in order to get the simulation of our model. Also, the formula [3.3] (with $T_d \wedge t$ instead of T_d) was used to calculate the total (discounted to the moment $t = 0$) dividend payment up to t.

The simulation of the two processes (Y_t itself and the process $V_t(x, b)$ of dividend payments up to t) is shown in Figure 3.2 (note that the upper picture is a magnification of the lower left corner of the lower picture).

The horizontal line (which is close to zero) denotes the dividend barrier $b = 10$, the blue curve fluctuating around it is Y_t.

The upper horizontal line shows the expectation of the total dividend payment, the vertical yellow line marking the time of Parisian ruin. It is clear that $V_t(x, b)$ (the yellow step function) is close to $\mathsf{E}V(x, b)$ at time $t = T_d$ of Parisian ruin.

The parameters here are as follows $\lambda = 5.4$, $c = 10.0$, $\beta = 1.0$, $x = 0.0$, $d = 0.5$, $h = 0.3$, $r = 0.01$.

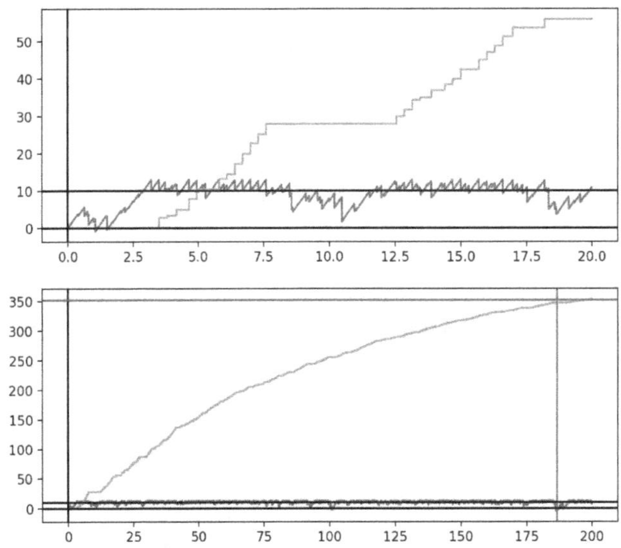

Figure 3.2. *Simulation of the process Y_t and dividends $V_t(x, b)$. For a color version of this figure, see www.iste.co.uk/dimotikalis/analysis2.zip*

3.3. Model 2

To emphasize the fruitfulness of cost approach, below we investigate the model with dividend payments, dual to the classical Cramér–Lundberg insurance model. That means, the company surplus (or capital) $U(t)$ at time t, without dividends, is described by the following relation:

$$U(t) = u - c_1 t + S(t),$$

where u is the initial surplus, c_1 is the expenses rate, and the last term $S(t) = \sum_{n=1}^{N(t)} Y_n$ represents the company profit. It is supposed that $N(t)$ is the Poisson process generated by a sequence of independent identically distributed (i.i.d.) non-negative random variables (r.v.'s) $\{T_n\}$ having exponential distribution with parameter λ. The sequence $\{Y_n\}$ does not depend on $N(t)$ and also consists of non-negative i.i.d. r.v.'s with distribution function $F(\cdot)$ and density $p(\cdot)$. Dual models (or models with negative claims) arise in life insurance (see the classical textbooks Bühlman (1970) and Gerber (1979)).

A pension fund or insurance company providing life annuities is a typical example. Here, c_1 is the rate of pension payments, and $S(t)$ is the fund gain up to time t caused by the deaths of their customers.

There exist other possible interpretations for a dual Cramér–Lundberg model. One can treat the surplus as the amount of capital of a business engaged in research and development (see, for example, Avanzi *et al.* (2007)). The company pays continuous expenses for research, and the occasional profit of random amounts (such as the award of a patent or a sudden increase in sales) arises according to the Poisson process. A similar model was used to model the functioning of a venture capital investment company in Bayraktar and Egami (2008).

Thus, the object of investigation is a dual Cramér–Lundberg model with dividends. If the dividend strategy with a constant barrier level $b > 0$ is applied, the surplus and aggregate dividends have the form depicted by Figure 3.3. In other words, whenever the surplus exceeds the barrier, the excess is paid out immediately as a dividend.

Expected discounted dividends paid until ruin time T_u (the first time that the surplus becomes negative) are given by $V(x; b) = E(\int_0^{T_u} e^{-\delta t} dD(t))$. Here, $D(t)$ is the aggregate dividends paid up to time t and $\delta > 0$ is a constant discount rate. It was established in Avanzi *et al.* (2007) that a simple barrier strategy is optimal for the model under consideration, and it is possible to calculate the optimal barrier for some particular cases.

Here, we consider a threshold strategy. The modified process (with dividends paid at the rate α after crossing the threshold b) has the form

$$U(t) = u - ct + S(t), \quad t \geq 0, \text{for } u \geq 0,$$

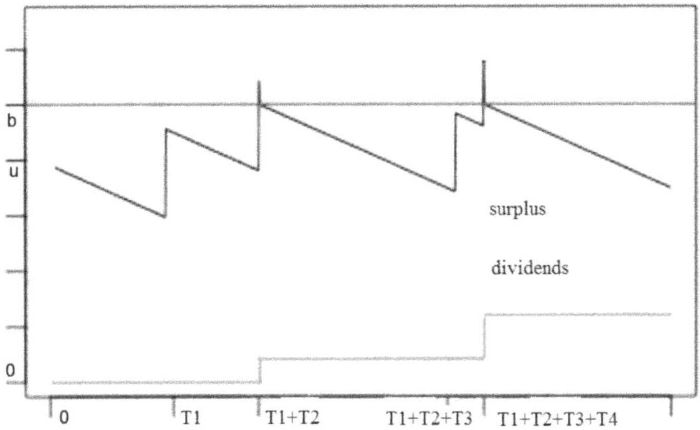

Figure 3.3. *Surplus and dividends under barrier strategy. For a color version of this figure, see www.iste.co.uk/dimotikalis/analysis2.zip*

where $c = \begin{cases} c_1, \ u \leq b; \\ c_2 = c_1 + \alpha, \ u > b. \end{cases}$

Thus, the dividends amount until the ruin time T_u is given by $D(b) = \alpha \int\limits_0^{T_u} e^{-\delta t} I(U(t) > b) dt$, whereas expected discounted dividends $V(u; b) = E[D(b)|U(0) = u]$ (see Figure 3.4).

THEOREM 3.4.– *For a threshold strategy (with level b)*

$$V(u; b) = \begin{cases} V_1(u), u \leq b, \\ V_2(u), u > b, \end{cases}$$

satisfies the system of integro-differential equations

$$\begin{cases} V(u; b) = 0, u = 0, \\ (\lambda + \delta) V_2(u) + c_2 V_2'(u) = \lambda \int\limits_0^{\infty} V_2(u + y) dF(y) + \alpha, u > b, \\ (\lambda + \delta) V_1(u) + c_1 V_1'(u) = \lambda \int\limits_0^{b-u} V_1(u + y) dF(y) + \\ + \lambda \int\limits_{b-u}^{\infty} V_2(u + y) dF(y), 0 < u \leq b. \end{cases}$$

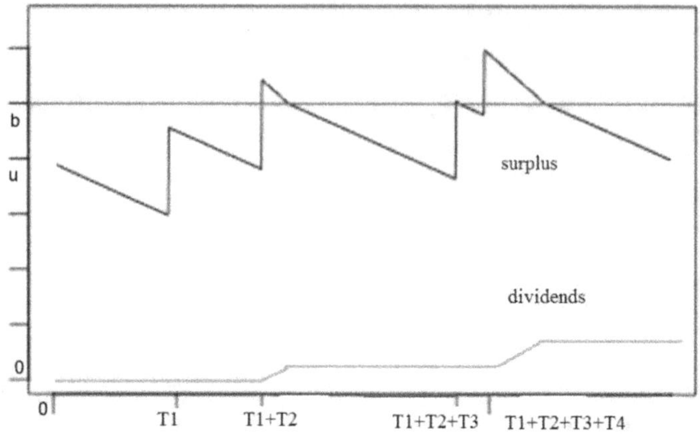

Figure 3.4. *Surplus and dividends under threshold strategy. For a color version of this figure, see www.iste.co.uk/dimotikalis/analysis2.zip*

THEOREM 3.5.– *For the exponential jumps, namely,* $p(y) = \beta e^{-\beta y}, y > 0$, *the expected discounted dividends*

$$V(u;b) = \begin{cases} A(e^{y_1 u} - e^{y_2 u}), 0 \le u \le b, \\ Ce^{x_1 u} + \frac{\alpha}{\delta}, u > b, \end{cases} \qquad [3.6]$$

where

$$A = \frac{-\alpha x_1}{\beta\delta} \cdot \frac{(\beta - y_2)(\beta - y_1)}{(y_2 - x_1)(\beta - y_1)e^{y_2 b} - (y_1 - x_1)(\beta - y_2)e^{y_1 b}} > 0,$$

$$C = -\frac{\alpha(\beta - x_1)}{\beta b} \cdot \frac{y_2(\beta - y_1)e^{y_2}b - y_1(\beta - y_1)e^{y_1 b}}{(y_2 - x_1)(\beta - y_1)e^{y_2 b} - (y_1 - x_1)(\beta - y_2)e^{y_1 b}} e^{-x_1 b} < 0.$$

To prove this result, we apply the operator $(\frac{d}{du} - \beta)$ to both sides of the integro-differential equations obtained in Theorem 3.4. This method was used, for example, in Avanzi *et al*. (2007).

Hence, for $u > b$, we get

$$c_2 V_2''(u) + (\lambda + \delta - \beta c_2)V_2'(u) - \delta\beta V_2(u) = -\beta\alpha.$$

Its characteristic equation has two roots $x_1 < 0, x_2 > 0$.

In the same way, we proceed in the case $0 < u \leq b$ and obtain the differential equation

$$c_1 V_1''(u) + (\lambda + \delta - \beta c_1) V_1'(u) - \delta \beta V_1(u) = 0.$$

Its characteristic equation also has two roots $y_1 < 0, y_2 > 0$. Solving two differential equations of the second order, we obtain the stated result.

Now, it is possible to obtain the optimal threshold.

THEOREM 3.6.– *The optimal threshold is given by*

$$b^* = \frac{1}{y_2 - y_1} \ln \frac{(y_1 - x_1)(\beta - y_2) y_1}{(y_2 - x_1)(\beta - y_1) y_2}.$$

In order to prove this result, we maximize the expressions of $V(u, b)$ for $0 \leq u \leq b$ and for $u > b$ separately, and find that the values of b^* providing the maximum coincide.

3.4. Conclusion and further results

First, we investigated the barrier dividend strategy with the Parisian implementation delay. In contrast to Dassios and Wu (2009), the expected discounted dividend payment until the Parisian ruin was considered as the objective function. As its explicit form is very complicated, we began by simulating the surplus process. It has shown us that in certain conditions, the optimal barrier is unique and does not depend on the initial surplus. An algorithm for its calculation was also obtained. The next step is parameter estimation. Due to restriction on the paper length, it is omitted. Moreover, under the condition of positive unit time profit, it is easy to see that for the process without dividends $X(t)/t \to \infty$, as $t \to \infty$. Therefore, it is interesting to study the distribution of recovery time $\eta = \sup\{t > 0 : X(t) < 0\}$. It is also necessary to carry out the sensitivity analysis and establish the stability conditions.

For the dual process with dividend threshold strategy, we only formulated the obtained results without the proofs. The next steps are the study of Parisian ruins (instead of the usual ones), the investigation of the optimal policy dependence on the gains distribution in terms of probability metrics, as well as the parameter estimation.

3.5. Acknowledgments

This research was partially supported by the RFBR grant 20-01-00487.

Many thanks to the anonymous reviewer for reading the chapter and making suggestions to improve the presentation.

3.6. References

Albrecher, H. and Thonhauser, S. (2009). Optimality results for dividend problems in insurance. *RESCAM Rev. R. Acad. Cien. Serie A. Mat.*, 103(2), 295–320.

Avanzi, B. (2009). Strategies for dividend distribution: A review. *N. Am. Actuar. J.*, 13(2), 217–251.

Avanzi, B., Gerber, H.U., Shiu, E.S.W. (2007). Optimal dividends in the dual model. *Insur. Math. Econ.*, 41(1), 111–123.

Azcue, P. and Muler, N. (2005). Optimal reinsurance and dividend distribution policies in the Cramér-Lundberg model. *Math. Finance*, 15(2), 261–308.

Bayraktar, E. and Egami, M. (2008). Optimizing venture capital investment in a jump diffusion model. *Math. Methods Oper. Res.*, 67(1), 21–42.

Bühlman, H. (1970). *Mathematical Methods in Risk Theory*. Springer-Verlag, Heidelberg.

Bulinskaya, E. (2017a). New research directions in modern actuarial sciences. In *Modern Problems of Stochastic Analysis and Statistics – Selected Contributions in Honor of Valentin Konakov*, Panov, V. (ed.). Springer, Cham.

Bulinskaya, E.V. (2017b). Cost approach versus reliability. *Proceedings of International Conference DCCN-2017*, Technosphera, Moscow.

Bulinskaya, E.V. and Shigida, B.I. (2018). Sensitivity analysis of some applied probability models (in Russian). *Fundam. Appl. Math.*, 22(3), 19–34.

Bulinskaya, E.V. and Shigida, I.B. (2019). Modeling and asymptotic analysis of insurance company performance. *Communications in Statistics – Simulation and Computation* [Online]. DOI: 10.1080/03610918.2019.1612911.

Dassios, A. and Wu, S. (2009). On barrier strategy dividends with Parisian implementation delay for classical surplus processes. *Insur. Math. Econ.*, 45, 195–202.

Drekic, S., Woo, J.-K., Xu, R. (2018). A threshold-based risk process with a waiting period to pay dividends. *J. Ind. Manag. Optim.*, 14(3), 1179–2001.

de Finetti, B. (1957). Su un'impostazione alternativa della teoria collettiva del rischio. *Transactions of the XV-th International Congress of Actuaries*, 2, 433–443.

Gerber, H.U. (1979). *An Introduction to Mathematical Risk Theory*. Huebner Foundation, Philadelphia.

Mikosch, T. (2006). *Non-life Insurance Mathematics*. Springer-Verlag, Berlin.

Sethi, S.P., Derzko, N.A., Lehoczky, J. (1991). A stochastic extension of Miller-Modigliany framework. *Math. Finance*, 1, 57–76.

4

Introduction of Reserves in Self-adjusting Steering of Parameters of a Pay-As-You-Go Pension Plan

The demographic trend of pension funds in Morocco (increased longevity combined with a drop in birth rates) and the situation of the labor market (a large share of the informal sector in employment) are a major challenge for the future of pay-as-you-go pension schemes in Morocco.

The mandatory Moroccan pension system operates in provisioned distribution and is financed by defined benefits. In the past, the surplus situation of the various plans has allowed them to accumulate significant financial reserves. In order to adapt to the structural challenges related to shortfalls in defined benefit management, several parametric reforms have been carried out, each time in order to postpone the date of exhaustion of reserves. Projections show that future parametric reforms will be unsustainable in terms of contribution rates or career extensions. Thus, it appears that a structural overhaul of the Moroccan pension system is more than necessary. We are going to propose to restructure the current system into a new system working through retirement points and piloted using the Musgrave rule.

4.1. Introduction

The defined benefit pension system in Morocco suffers from demographic problems related first of all to the increase in life expectancy at birth, meaning that retirees are living longer and this increase in life expectancy is combined with a fall in the total fertility rate (average number of children per woman), which in Morocco has

---

Chapter written by Keivan DIAKITE, Abderrahim OULIDI and Pierre DEVOLDER.

gone from 7.7 in 1962 to 2.49 in 2016 and is forecast to decrease further in the coming years; as long as it remains below the renewal threshold of the population, there will be fewer active workers to finance more and more pensions; thus, in 2016, the number of active workers for a pensioner barely reaches 2.23, while this ratio was 6 in 2000.

Added to this demographic problem is an unfavorable economic situation. The sustainability of the distribution lies in the fact that the contributions are greater than the wage bill is, but the unemployment situation is significant (10.6 %) in addition to the fact that the nature of employment is not the same as before. The fixed wage is tending to disappear in favor of entrepreneurship, so the number of employee contributors is continuing to decline, which weakens the regime. In addition, the informal sector accounts for 40% of employment in 2016 (see the CESE annual report 2017).

Faced with this challenge of longevity, many countries around the world have started (or will begin) reforms of their mandatory pension scheme. These reforms maintained defined benefits and adjusted various parameters (postponement of retirement age, reduction of benefit rates, hardening of anticipation conditions in the event of early retirement...). We are talking about parametric reforms. These punctual measures can, up to a certain point, restore a certain viability in the short term, but in the face of the scale of the actual challenges, they are obsolete.

This chapter focuses on the transformation of the current pay-as-you-go (PAYG) system with defined benefits into a point-managed pension system and the introduction of a rule of automatic piloting of the different parameters of the regime over time: the Musgrave rule. To do this, we will briefly present the architecture of the Moroccan pension system as a whole before focusing on the largest public pension fund by presenting its characteristics and parameters.

In the second part of the chapter, we present the theoretical framework of the Musgrave rule in the management and control of the regime in point as well as the effect of the introduction on the extinction date of the reserves. We will then simulate the transformation of the fund into a point-managed plan by applying the Musgrave rule and the introduction of reserves. Finally, we compare the current system with the new simulated system by measuring the impact of this transformation on the level of benefits and contributions through contribution rates and replacement rates.

4.2. The pension system

The current Moroccan system is based exclusively on a contributory funded PAYG pillar. It consists of three compulsory basic schemes and a conventional supplementary scheme, differentiated by a professional category. Non-salaried workers, i.e. traders and craftsmen, the liberal professions, farmers and fishermen as well as mobile workers, cannot join any scheme at present. The system covers 40.9% of the employed labor force in 2018.

Our study was focused on the Moroccan pension fund (CMR); we present in what will follow the characteristics of the fund and the different parameters.

The CMR public scheme is compulsory for three categories of employees: the civil and military personnel of the State, permanent and trainee agents of local authorities and the staff of public institutions.

The plan is funded on a PAYG basis. The contribution rate is set at 28% of base salary, bonuses and other allowances. The contribution rate is equally shared between employees and the state employer. The plan is based on the principle of the laddered premium which sets an equilibrium contribution rate for a minimum period of 10 years.

The pension is calculated as such:

$$P = N \times A \times SR \qquad [4.1]$$

where:

- N is the number of years contributed;
- A is the annuity rate;
- SR is the reference salary.

Before 2016, the annuity rate was 2.5%; the reference salary was the last salary, and the legal retirement age was 60 years; this caused the scheme to be much too generous. It offered for 40 years of contribution a replacement rate of 100%. The parametric reform tried to correct this generosity. Thus, the maximum contribution period within the fund is 40 years and the reference salary for the calculation of the pension is now the average of the last eight (8) earnings preceding the date of retirement. The annuity rate is now 2%. The fund offers a maximum of 80% replacement rate for a complete career. The legal retirement age is 63 years.

The fund's surplus position in the past has allowed it to accumulate significant reserves. Today, these reserves make it possible to fill the technical deficit. However, the evolution of the declining population ratio has accelerated the depletion of these reserves. The evolution of the demographic ratio is presented here.

The main assumption used in this projection is the replacement of the workers, and their number remains the same on the projection horizon. We used a deterministic projection for the retiree population.

We will use five professional categories representing career trajectories for our simulations; we present the average wages by age for the following categories, the wages are in Moroccan Dirham:

– administrators;

– engineers;

– grade A professors;

– secretaries;

– grade B teachers.

The selected categories are representative of wage developments within the fund.

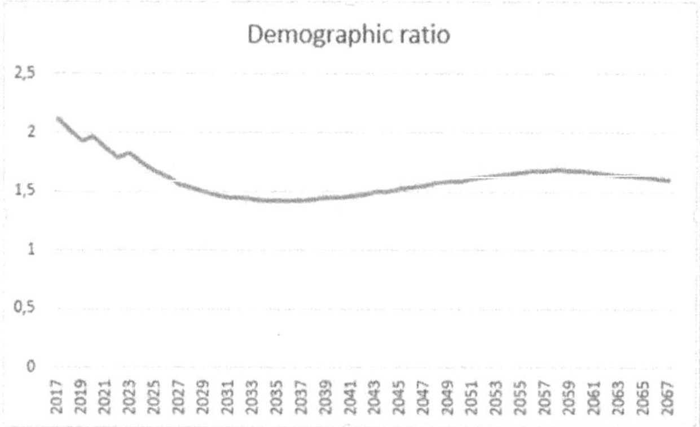

Figure 4.1. *Projection of the demographic ratio of CMR*

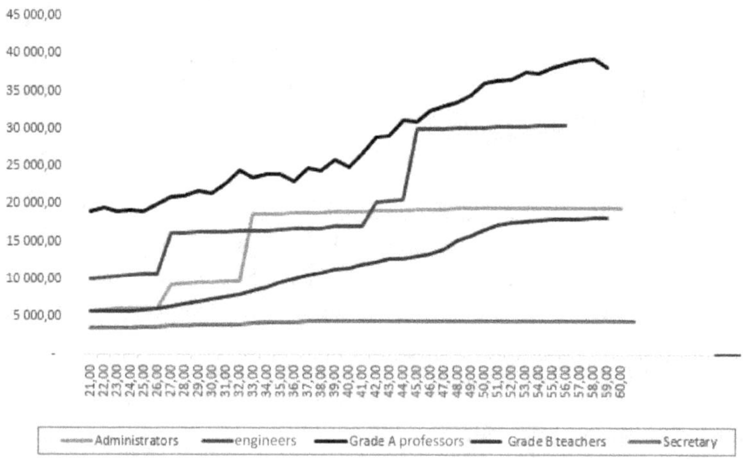

Figure 4.2. *Evolution of wage trajectory by age. For a color version of this figure, see www.iste.co.uk/dimotikalis/analysis2.zip*

The secondary school teacher career trajectories (grade B teachers), administrators, higher education professors (grade A professors) and state engineers have the same increasing pace depending on the length of their career. People in these occupations have long careers in the scheme (between 35 and 40 years). Salaries change gradually. Secretaries have "flat" careers and benefit from a very slight evolution. We show the distribution of the workforce in those categories.

	Administrators	Engineers	A professors	B teachers	Secretaries
Distribution	24%	8%	3%	63%	2%

Table 4.1. *Distribution of the workforce between the categories*

Administrators and grade B teachers represent more than 80% of the members of the scheme.

Using the current parameters, we are going to simulate on these trajectories the contributions throughout their career, and determine the replacement rates associated with the average contribution period in a year for each trajectory as well as the ratio between expected benefits and contribution through the career.

	Administrators	Engineers	A professors	B teachers	Secretaries
Average contribution period	40	35	38	38	40
Replacement rate	80%	75%	76%	75%	80%
Benefits/contributions	0.94	1.20	1.03	1.25	0.81

Table 4.2. *Replacement rates and ratio between benefits and contributions*

The individuals having contributed 40 years have a maximal replacement rate. The current system only rewards the long haul. The type of career has no influence on replacement rates. As a result, the ratio of benefits paid on contributions is better for careers with a good valuation, and careers with a low reevaluation are disadvantageous. The scheme pays on average for each monetary unit contributed 1.15 in return.

After describing the functioning of the Moroccan Pension Fund, we will present in what follows the theoretical framework of the rule of piloting the new regime that we will put in place.

4.3. Theoretical framework of the Musgrave rule

When the demographic indicators deteriorate, depending on the management method, parameters such as contribution and replacement rates adjust to compensate for the decrease. We will present this mechanism when the fund is managed in defined benefits and then we will introduce a new management mode driven by the Musgrave rule.

We model the demographic risk by assuming a stable state (noted in state 1) composed of representative agents (same salary and same career) receiving a pension based on a replacement rate δ_1 and a contribution rate π_1; the dependency ratio (ratio of the number of retirees to number of contributors) is D_1. The balance of the regime is characterized by the following system, where P_t is the average pension paid and S_t is the average salary:

– Budget equation : $D_1 \cdot P_t = \pi_1 \cdot S_t$

– Pension equation : $P_t = \delta_1 \cdot S_t$

Equilibrium is obtained when:

$$\pi_1 = D_1 \cdot \delta_1 \quad [4.2]$$

We suppose now that the system moves to another stage characterized by another dependence ratio $D_1 \to D_2$. We assume that $D_2 > D_1$. We find the contribution rate and the replacement rate in the second state linked by:

$$\pi_2 = D_2 \cdot \delta_2 \quad [4.3]$$

In a DB framework, there is an absolute guarantee for the retirees (fixed replacement rate), and the contributors must support the risks

$$\Rightarrow \delta_2 = \delta_1 = \delta$$

$$\pi_2 = \pi_1 \cdot \frac{D_2}{D_1} \quad [4.4]$$

The demographic changes taking place within the fund are one of the main causes of the depletion of reserves. The defined benefit system puts the burden on the workers (contribution rate). Thus, according to the projections, when the demographic ratio decreases to 1.37 by 2050, the contribution rate should grow from 28% to 40%. Such contribution rates are too burdensome for the contributors only.

Musgrave (1981) proposed another invariant leading to a form of sharing of the risk between the two generations. Let us define the Musgrave ratio as the ratio between the pension and the salary net of pension contributions:

$$M_1 = \frac{P}{S(1-\pi_1)} = \frac{\delta_1}{1-\pi_1} \quad [4.5]$$

Using the previous situation, when D_1 becomes D_2, we stabilize this coefficient:

$$M_1 = \frac{\delta_1}{1-\pi_1} = \frac{\delta_2}{1-\pi_2} = M_2 \quad [4.6]$$

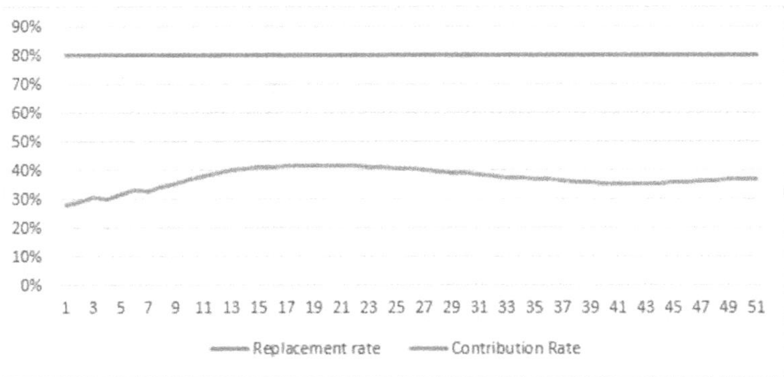

Figure 4.3. *Projection of contribution rates and replacement rate in DB. For a color version of this figure, see www.iste.co.uk/dimotikalis/analysis2.zip*

Using [4.2] and [4.3]:

$$\frac{\delta_1}{1-\pi_1} = \frac{\delta_2}{1-\pi_2} \Rightarrow \delta_2 = \delta_1 \cdot \frac{1-\pi_2}{1-\pi_1}$$

$$\delta_2 = \delta_1 \cdot \frac{1-D_2 \cdot \delta_2}{1-D_1 \cdot \delta_1}$$

We deduce δ_2 and arrive at:

$$\boxed{\delta_2 = \delta_1 \cdot \frac{1}{1+\delta_1(D_2-D_1)}} \qquad [4.7]$$

The contribution rate in the second state is determined in the same way and we have:

$$\boxed{\pi_2 = \pi_1 \cdot \frac{D_2}{D_1+\pi_1(D_2-D_1)}} \qquad [4.8]$$

We present here the evolution of the contribution rate and the replacement rate using the Musgrave rule.

It can be seen that the wage contribution that makes it possible to balance the plan from an actuarial point of view goes from 28% to 35% when the dependency ratio is the most deteriorated. The contribution rate drops to 69% at this period. The evolution of the rates follows the tendency of the dependence ratio; when it deteriorates, the contribution rates rise and the replacement rate becomes lower (see Figure 4.4). The deterioration of the demographic ratio is absorbed by both the active and the retired in the simulation. In the next section, we will apply this piloting rule on the parameters and transform the scheme.

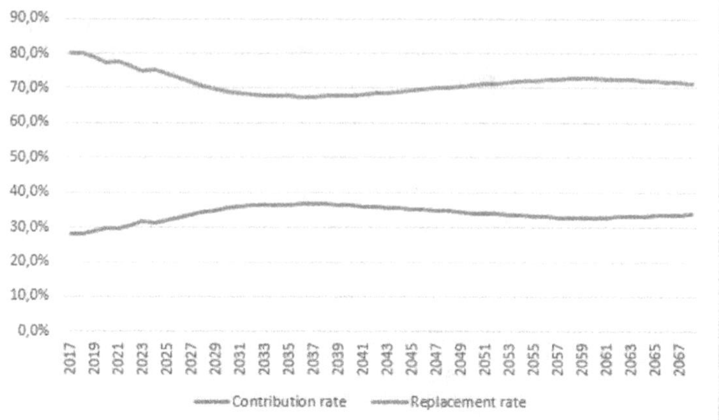

Figure 4.4. *Projection of contribution rates and replacement rate under the Musgrave rule. For a color version of this figure, see www.iste.co.uk/dimotikalis/analysis2.zip*

4.4. Transformation of the retirement fund

After showing the problems related to defined benefit management on pensions, we have proposed the transformation on the current DB system of the CMR into a new one financed in PAYG; the benefits are calculated using a point system and the risks are shared between the retiree and the contributors. The system is described as follows.

Each year, the payment by the affiliate of his/her contribution entitles him/her to a certain number of retirement points. The number of points given is the ratio between the salary of the affiliate and an identical reference salary for all, called the acquisition value of the point.

The monetary counterpart of these points is only known on the liquidation date, depending on the value of service of the point on that date.

The number of points earned at the time of retirement is the sum of the points earned during the career. The pension at retirement age is given by the formula:

$$P = N_T \cdot V_T \cdot \sigma_T \qquad [4.9]$$

where N_T represents the number of points earned at retirement age T, V_T is the value of the point and σ_T is an actuarial coefficient that depends on the length of career and the generation.

In order to determine the liquidation value of the point, we consider an individual who has contributed for a period N with a salary each year equaling the reference salary. The actuarial coefficient equals 1 for this individual. The amount of the pension P_T can be written according to the replacement rate δ and the reference wage S_T^r

$$P_T = \delta \cdot S_T^r \tag{4.10}$$

and according to the number of points and the value of one point

$$P_T = N \cdot V_T \tag{4.11}$$

we deduce the value of the point:

$$\boxed{V_T = \frac{\delta \cdot S_T^r}{N}} \tag{4.12}$$

The point system presented in this section is a very flexible architecture and can be modeled using various calibrations. We can fix the value of the point and adapt it automatically through the evolution of the replacement rate.

We simulate through the five wage trajectories presented in section 4.2 the transformation of the current scheme into a new one managed with the point system we have just introduced. For individuals joining the plan today, we calculate their pension entitlements and the ratio of the present value between the benefits and the contributions. We will compare those indicators in the current system and after transformation.

	Administrators	Engineers	A professors	B teachers	Secretaries
Replacement rate	78%	55%	69%	60%	99%
Benefit/contribution ratio	0.9199	1.0110	0.8843	0.9832	0.8858

Table 4.3. *Replacement rates and contribution ratio in the new system*

Indexing the pension in relation to the evolution of the average salary of the scheme has the immediate effect of improving the value of the pension for "flat" trajectories; trajectories with evolution rates higher than the evolution of the average wage have replacement rates lower than the target replacement rate of the scheme; the redistribution of wealth in the scheme is done more uniformly across the types of trajectories.

Replacement rates are not capped; thus, trajectories with pay decreases at the end of the career have better replacement rates; they are also better for long careers (contribution period greater than the reference period) and do not penalize the fact of having a flat salary evolution. The scheme is more generous for trajectories having little revaluation throughout the career; this is the case for secretaries.

The second indicator measures the performance of the scheme; for each monetary unit paid in the form of a contribution, the Moroccan Pension Fund pays an average of

0.96. It is 0.19 less than the current system, and the system is on average less generous. The new system benefits contributors with wage developments that are lower than the average wage in the scheme, so there is a different distribution of wealth in the scheme.

After transforming the pension plan, we are interested in the impact on the level of the reserves as well as on the horizon of viability. Here is how we model the reserves:

$$Reserves_{t+1} = Reserves_t \cdot (1 + r_t) + Contributions_t - Prestations_t \qquad [4.13]$$

The rate of return is supposed to be constant. This rate corresponds to the average value observed over the last 10 years. The increase in contribution rates and the decline in replacement rates should slow down the rate of exhaustion of the fund, as the level of implicit debt is very high because the fund has operated in the past with generous parameters.

Before transforming the scheme, we present a projection of the reserve fund. There is no adjustment.

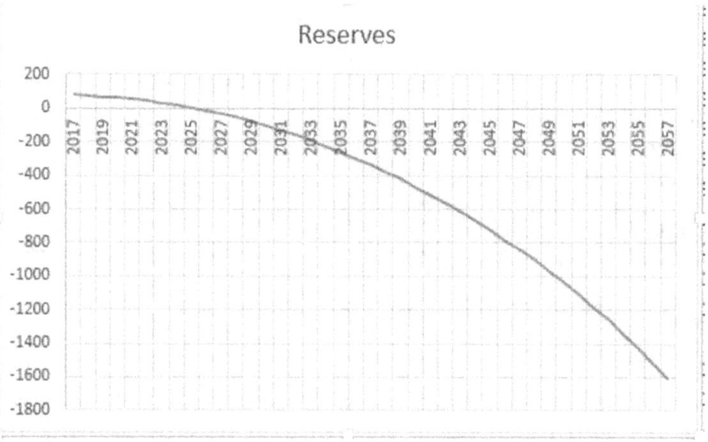

Figure 4.5. *Projection of the reserve fund without adjustment*

We observe that with the current operating parameters of the fund, it is possible to maintain a positive level of reserves only until 2027. The management is not financially sustainable in the long term. The deficit continues to grow until the horizon of projection.

After transformation, we introduce the Musgrave rule that allows us to control the level of contributions and benefits, as well as measure the impact on the level of reserves in the medium and long terms.

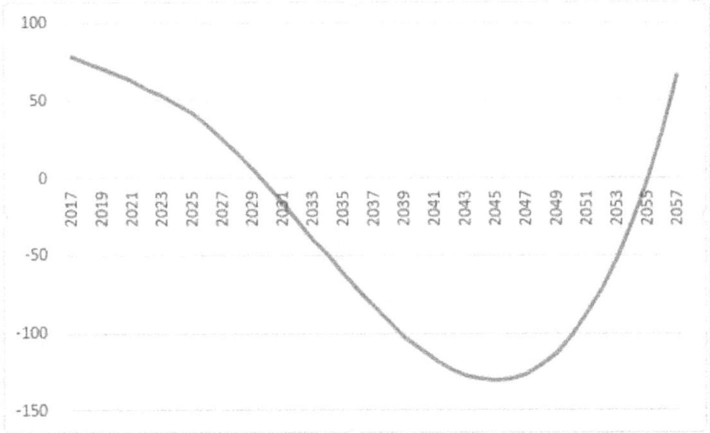

Figure 4.6. *Projection of the reserve fund with adjustment*

The transformation to a point-based system pushes the date of exhaustion of the reserves to 2031. This is a gain of four years of operation of the regime in addition compared to the management in defined benefits. The system's pricing should be able to balance the technical result in the medium-term regime; at the end of the projection horizon, the level of the reserves is positive again due to the piloting of the contribution and replacement rates with the Musgrave rule.

4.5. Conclusion

Facing the failing situation of PAYG financed pension plans managed in defined benefits in Morocco, we examined in this chapter a management model, the point system and a steering mechanism of the plan that would make it possible to overcome the shortcomings of the current system. Our purpose was to determine whether the new scheme is financially sustainable and its impact on the standard of living of the contributors and retirees. We first drew a portrait of the current situation of the Moroccan retirement system. This analysis allowed us to identify the diverging parameters from one fund to another as well as the problems related to the defined benefit pension management method. We also highlighted the actions taken to solve these problems. Then, we presented the theoretical model of the points and Musgrave's rule to control the value of the point as well as contribution rates and equity in the distribution of capital through active workers and retirees, also across generations. Under pressure from the deterioration of the demographic ratio, the solutions to maintain thc PAYG pension system are becoming fewer and fewer. The point system allowed this load to be distributed with equity.

In this chapter, we have considered a deterministic approach to model the evolution of the demography and the rates of return. Future extensions will examine stochastic models for the rate of return of reserves.

4.6. References

Blanchard, M. (2017). Pilotage et gestion d'un régime de retraite et impact sur sa situation financière suite au decret. Report, Institut des actuaires, 2017-887.

Caisse marocaine des retraite (2016). Activity report.

Comission 2020–2040 (2014). Un contrat social performant et fiable [Online]. Available: http://pension2040.belgium.be/fr/.

Conseil Economique Social et Environmental (2017). Annual report, France.

Cour des comptes (2018). Rapport sur caisse marocaine des retraites. Report, Cour des comptes, Morocco.

Devolder, P. (2010). Perspectives pour nos régimes de pension légale. *Revue belge de sécurité sociale*, 4, 597–614.

Devolder, P. (2015). Pension reform in Belgium, a new points system between DB and DC. [Online]. Available: http://www.actuaries.org/oslo2015/papers/PBSS-Devolder.pdf.

Musgrave, R. (1981). A reappraisal of social security finance. *Social Security Financing*. Cambridge, MIT, 89–127.

Palier, B. (2012). *La réforme des retraites*. Presses Universitaires de France, Paris.

5

Forecasting Stochastic Volatility for Exchange Rates using EWMA

In risk management, foreign investors or multinational corporations are highly interested in knowing how volatile a currency is in order to hedge risk. In this chapter, using daily exchange rates and the exponential weighted moving average (EWMA) model, we perform volatility forecasting. We will investigate how the use of the available time series affects the forecasting, i.e. how reliable our forecasting is depending on the period of available data used. We will also test the effects of the decay factor appearing in the model used on the forecasts. The results show that, for the data used, it is optimal to use a larger value of the decay factor and also, for longer out-of-sample periods, the forecasts get closer to reality.

5.1. Introduction

When you move into a country that uses a different currency than yours, you need to change your currency by buying that of the country you are moving to. The rate at which you buy is called the exchange rate (Hull 2006), which is greater or less than one, depending on whether your currency is lower or greater in value than the currency you are buying.

Considering, for example, the currency of Sweden (the Swedish Krona: SEK) and the currency of Rwanda (the Rwandan Franc: RWF), the exchange rate for the pair SEK/RWF is greater than 1 because 1 SEK is nowadays nearly equivalent to 100 RWF. When the exchange rate is equal to one, both the considered currencies are equal in power, but this rarely happens.

Chapter written by Jean-Paul MURARA, Anatoliy MALYARENKO, Milica RANCIC and Sergei SILVESTROV (MSC 2020 Classification: 62P05).

In the following, we will use S_t to denote the exchange rate between two currencies at a certain date t. The change in the returns of an exchange rate for a given period is known as *volatility* (Hull 2006). When hedging risk, risk managers have interest in knowing how volatile a specified currency pair is. Thus, financial engineering brings different techniques on how to model volatility in financial markets. By modeling volatility here, we mean that we are trying to derive a model that can be used to best forecast volatility for future returns.

For a given currency pair, let us consider the return R_t:

$$R_t = \frac{S_t - S_{t-1}}{S_{t-1}} = \frac{S_t}{S_{t-1}} - 1.$$

The corresponding logarithmic return (continuous compounded return) is:

$$r_t = \ln(1 + R_t) = \ln\left(\frac{S_t}{S_{t-1}}\right).$$

EXAMPLE 5.1.– *The USD/RWF pair, on December 27 and 28, 2018, was* 878.9566 *and* 879.0868, *respectively. In this case,* $S_{t-1} = 878.9566$ *and* $S_t = 879.0868$. *Thus, the corresponding return is*

$$R_t = \frac{879.0868 - 878.9566}{878.9566} = \frac{0.1302}{878.9566} = 1.4814 \times 10^{-4}$$

and its logarithmic return is:

$$r_t = \ln(1 + 1.4814 \times 10^{-4}) = 1.4813 \times 10^{-4}.$$

The exponential weighted moving average (EWMA) is one of the models used to estimate the volatility in financial returns. It can be written as follows:

$$\sigma_t^2 = \lambda\sigma_{t-1}^2 + (1 - \lambda)r_{t-1}^2 \qquad [5.1]$$

where $\lambda \in [0, 1]$ is the decay factor, σ_t^2 is the variance at time t and r_t is the log-return at time t.

In this chapter, two important questions are investigated, the first being the best value of the decay factor in the EWMA model when forecasting volatility of exchange rates and the second being the optimal out-of-sample forecasting period. Before reviewing the EWMA model used (Winters 1960; J.P. Morgan 1996; Bollen 2015), let us have a look at the data analyzed in this chapter.

5.2. Data

We deal with five currencies: EUR (Euro), USD (US Dollar), SEK (Swedish Krona), KES (Kenyan Shilling) and RWF (Rwandan Franc). These data have been collected from the website of the National Bank of Rwanda (BNR) (2019). The collected data are four time series of daily exchange rates that cover the period from January 1, 1995 up to December 31, 2018. Because some information related to EUR and SEK is missing for early years, we have optimized by equalizing the ranges and then have chosen to work with the period starting from January 1, 2003 up to December 31, 2018.

The four currency pair series used (EUR/RWF, KES/RWF, SEK/RWF and USD/RWF) have 5,844 observations each, which makes 23,376 observations in total.

One of the novelties in this chapter is a form of extrapolation in the collected data. The missing values for weekends and holidays have been filled in by considering their corresponding previous values. We have introduced this method based on the fact that the returns for values around the missing values are stationary. This allows us to consider a year with 365.25 trading days instead of the usual 252 generally used. The descriptive statistics are given in Table 5.1.

	EUR/RWF	**USD/RWF**	**SEK/RWF**	**KES/RWF**
Size	5,844	5,844	5,844	5,844
Minimum	533.2172	509.1101	58.2864	5.7595
Maximum	1,0652	879.1009	108.0807	8.8294
Range	531.9892	369.9908	49.7943	3.0699
Mean	803.4626	638.4323	85.8304	7.5951
St. dev.	104.4771	104.3494	10.9031	0.4832
Kurtosis	2.9081	2.6037	2.2614	3.4667
Skewness	0.1560	0.9853	-0.1927	0.0671

Table 5.1. *Descriptive statistics of raw data*

In Figure 5.1, we have plotted the normalized raw data. In this figure, the pair SEK/RWF has been multiplied by 10, while the pair KES/RWF has been multiplied by 100. This is to allow a better visualization of the data in the figure.

5.3. Empirical model

In 1994, J.P. Morgan, a financial services company (J.P. Morgan 1996), introduced some procedures to quantify financial risk in what has been called RiskMetrics. The EWMA volatility model has been added to the RiskMetrics in 1996 (J.P. Morgan 1996). From the generalized autoregressive conditional heteroscedasticity GARCH(1,1) model (Hull 2006), we have

$$\sigma_t^2 = \gamma + \beta\sigma_{t-1}^2 + \alpha r_{t-1}^2. \qquad [5.2]$$

While the EWMA model (Winters 1960) is given as:

$$\sigma_t^2 = \lambda\sigma_{t-1}^2 + (1-\lambda) r_{t-1}^2 \qquad [5.3]$$

which can also be written as:

$$\sigma_t^2 = \lambda^n \sigma_{t-n}^2 + (1-\lambda)\sum_{i=1}^{n} \lambda^{i-1} r_{t-i}^2, \qquad [5.4]$$

i.e. the EWMA is a particular case of GARCH(1,1) with $\gamma = 0$, $\beta = \lambda$ and $\alpha = 1 - \lambda$.

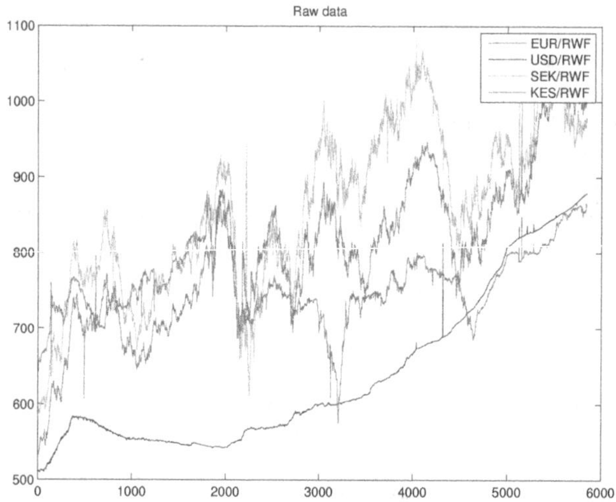

Figure 5.1. *Exchange rate data. Note that SEK/RWF (× 10) and KES/RWF (× 100). For a color version of this figure, see www.iste.co.uk/dimotikalis/analysis2.zip*

For large n, the term $\lambda^n \sigma_{t-n}^2$ tends to zero and this brings EWMA to the form:

$$\sigma_t^2 = (1-\lambda)\sum_{i=1}^{n} \lambda^{i-1} r_{t-i}^2. \qquad [5.5]$$

When using the EWMA model, the main goal is to estimate the next period or next day volatility in a time series and also to closely observe the way volatility changes (Andersen *et al*. 2005). The EWMA model uses two parameters: *time* and λ, which is related to the sensitivity of the forecasted volatility to the historical data.

The parameter λ satisfies: $0 < \lambda < 1$, and the RiskMetrics by J.P. Morgan (1996) suggests the use of $\lambda = 0.94$ for daily data and $\lambda = 0.97$ for monthly data. For better analysis, we choose to use $\lambda_1 = 0.97$, $\lambda_2 = 0.75$, $\lambda_3 = 0.50$ and $\lambda_4 = 0.25$, as suggested in Bollen (2015).

5.4. Exchange rate volatility forecasting

Let V_1 be the rolling historical volatility and V_2 be the EWMA volatility described by the following formulae:

$$V_1 = \sigma_{1,t} = \sqrt{\frac{365.25}{n-1}\sum_{i=1}^{n}(r_{t-i} - \bar{r}_t)^2} = \sqrt{\frac{365.25}{n-1}\sum_{i=1}^{n} r_{t-i}^2} \quad [5.6]$$

and

$$V_2 = \sigma_{2,t} = \sqrt{\lambda\sigma_{2,t-1}^2 + (1-\lambda)r_{t-1}^2} \quad [5.7]$$

where r_t is the logarithmic return for each pair and $\bar{r}_t$ is the related mean return with $t \in [1, T]$, $i \in [1, n]$ and $T = 5,843$, where n represents the window's size. Three values of n are used: $n = 7$, $n = 30$ and $n = 90$ for one-week, one-month and one-quarter window sizes, respectively.

We have chosen to work with 365.25 trading days instead of the common 252 because exchange bureaux operate each day of the year; thus, in this market, we need to consider all days of the year. Table 5.4 presents the descriptive statistics for different returns, and Figure 5.2 shows how the logarithmic returns behave.

	EUR/RWF	USD/RWF	SEK/RWF	KES/RWF
Size	5,843	5,843	5,843	5,843
Minimum	-0.1753	-0.1643	-0.2528	-0.1777
Maximum	0.1778	0.1647	0.2874	0.1545
Range	0.3531	0.3290	0.5402	0.3322
Mean	1.0829×10^{-4}	9.2550×10^{-5}	8.8752×10^{-5}	5.0016×10^{-5}
St. dev.	0.0074	0.0032	0.0095	0.0058
Kurtosis	225.8287	2.3790×10^{3}	272.9528	286.0107
Skewness	-0.5753	0.0800	1.4994	-1.5644

Table 5.2. *Descriptive statistics of logarithmic returns*

We compare the results of V_1 versus V_2 by observing two statistics, namely, the root mean square error (RMSE) and the mean absolute percentage error (MAPE), defined, respectively, as:

$$RMSE = \sqrt{\frac{1}{T}\sum_{t=1}^{T}(V_{1,t} - V_{2,t})^2} \quad ; \quad MAPE = \frac{1}{T}\sum_{t=1}^{T}\frac{|V_{1,t} - V_{2,t}|}{V_{1,t}}. \quad [5.8]$$

We apply four different values for the decay factor in the EWMA:

$\lambda_1 = 0.97$, $\lambda_2 = 0.75$, $\lambda_3 = 0.50$ and $\lambda_4 = 0.25$.

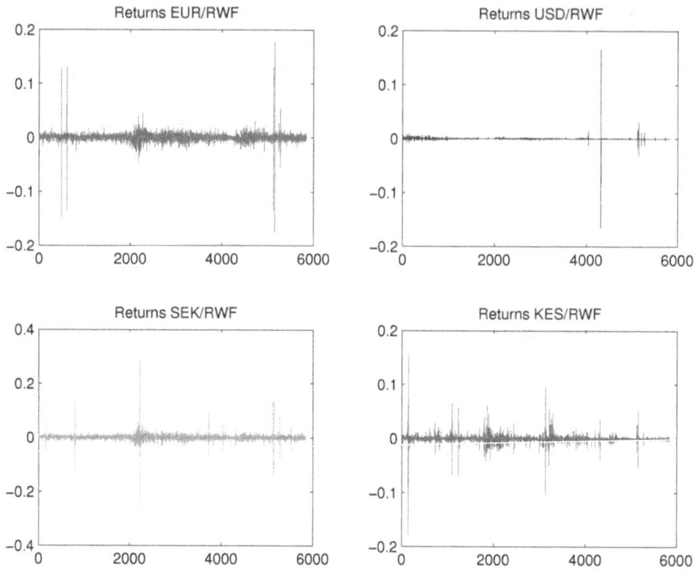

Figure 5.2. *Logarithmic returns. For a color version of this figure, see www.iste.co.uk/dimotikalis/analysis2.zip*

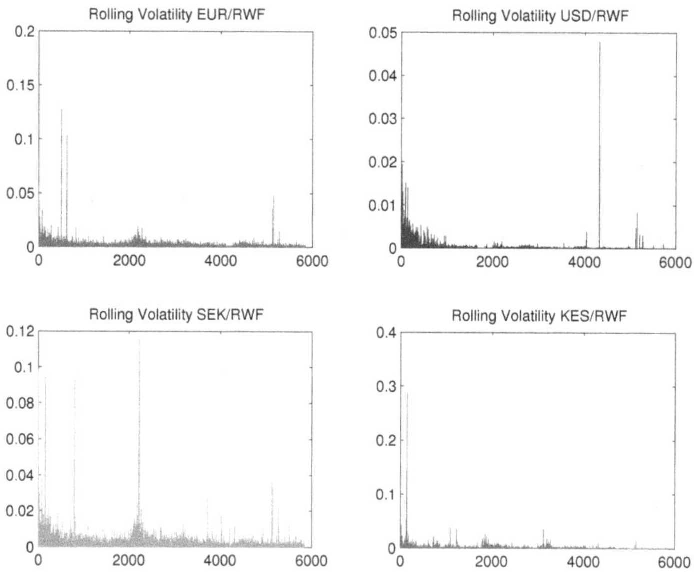

Figure 5.3. *Rolling volatility. For a color version of this figure, see www.iste.co.uk/dimotikalis/analysis2.zip*

The results obtained from the calculations of $RMSE$ and $MAPE$ give good results. The obtained values are given in Table 5.3. We remark that the value of the decay factor λ does not have any effect on the two types of errors calculated.

Week	EUR/RWF	USD/RWF	SEK/RWF	KES/RWF
RMSE (λ_1)	0.1732	0.0057	0.2339	0.1237
RMSE (λ_2)	0.1732	0.0057	0.2267	0.1198
RMSE (λ_3)	0.1731	0.0057	0.2255	0.1190
RMSE (λ_4)	0.1731	0.0057	0.2264	0.1192
MAPE (λ_1)	0.9999	0.9999	0.9770	0.9768
MAPE (λ_2)	0.9999	0.9999	0.9474	0.9456
MAPE (λ_3)	0.9998	0.9999	0.9424	0.9382
MAPE (λ_4)	0.9998	0.9999	0.9463	0.9397
Month	**EUR/RWF**	**USD/RWF**	**SEK/RWF**	**KES/RWF**
RMSE (λ_1)	0.3575	0.0144	0.4194	0.1748
RMSE (λ_2)	0.3540	0.0143	0.4148	0.1735
RMSE (λ_3)	0.3550	0.0143	0.4170	0.1742
RMSE (λ_4)	0.3567	0.0144	0.4197	0.1750
MAPE (λ_1)	0.9592	0.9595	0.9588	0.9593
MAPE (λ_2)	0.9500	0.9524	0.9491	0.9515
MAPE (λ_3)	0.9531	0.9555	0.9548	0.9553
MAPE (λ_4)	0.9576	0.9593	0.9612	0.9602
Term	**EUR/RWF**	**USD/RWF**	**SEK/RWF**	**KES/RWF**
RMSE (λ_1)	0.6379	0.0850	0.8632	0.2135
RMSE (λ_2)	0.6387	0.0871	0.8668	0.2127
RMSE (λ_3)	0.6401	0.0872	0.8702	0.2136
RMSE (λ_4)	0.6423	0.0873	0.8742	0.2147
MAPE (λ_1)	0.9614	0.9687	0.9617	0.9560
MAPE (λ_2)	0.9625	0.9899	0.9655	0.9515
MAPE (λ_3)	0.9645	0.9909	0.9692	0.9571
MAPE (λ_4)	0.9678	0.9922	0.9737	0.9615

Table 5.3. *Errors (RMSE and MAPE) for different decay factors λ_i and out-of-sample periods*

From Figures 5.4 to 5.6, we plot the real data versus forecasted ones for the three out-of-sample periods. The results in Figures 5.4–5.6, related to $\lambda_1 = 0.97$ and others not presented, show that the EWMA with a larger decay factor λ (i.e. closer to 1) is better for forecasting exchange rates. We can mention that, in the EWMA model with a large decay factor, the recent values have more weight than the non-recent ones on the forccasts (Andersen *et al.* 2005).

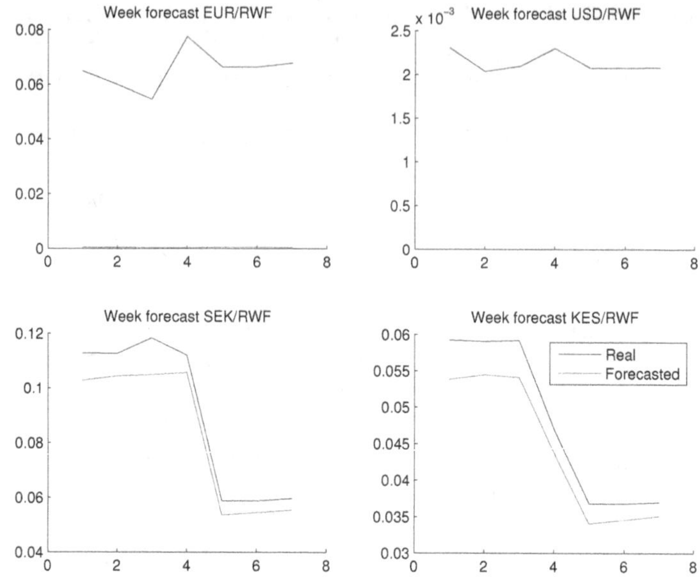

Figure 5.4. *One-week volatility forecasts with* $\lambda = 0.97$. *For a color version of this figure, see www.iste.co.uk/dimotikalis/analysis2.zip*

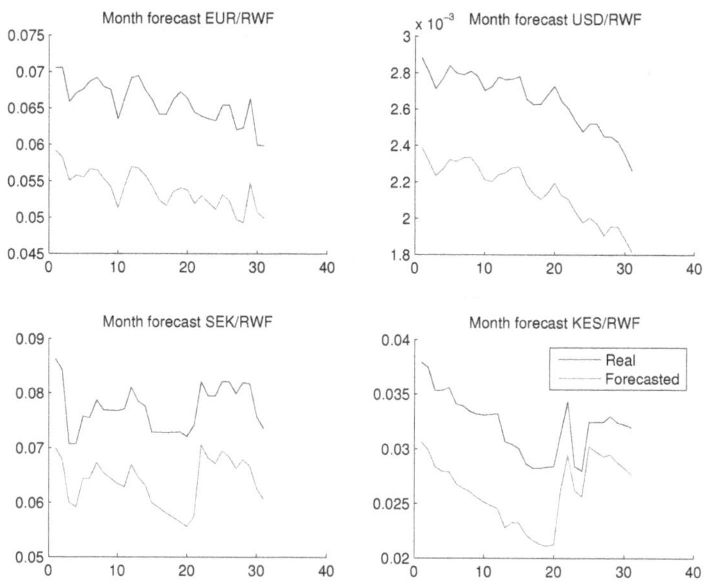

Figure 5.5. *One-month volatility forecasts with* $\lambda = 0.97$. *For a color version of this figure, see www.iste.co.uk/dimotikalis/analysis2.zip*

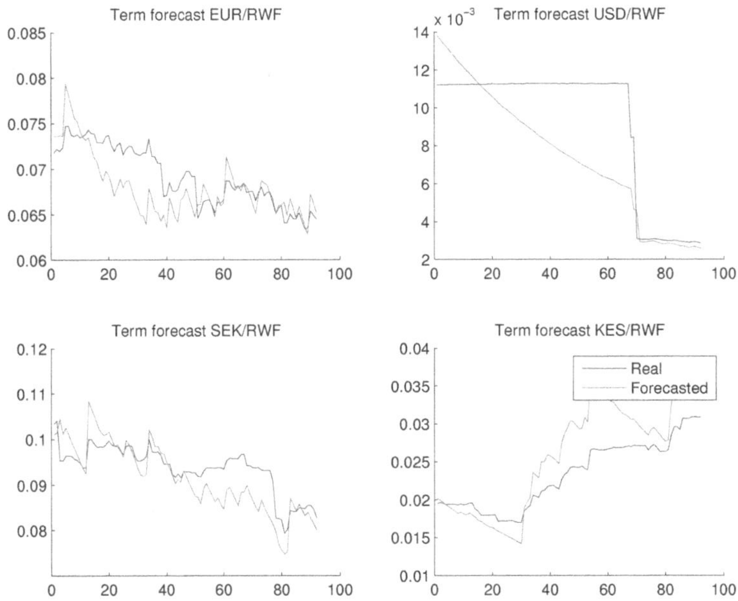

Figure 5.6. *One-quarter volatility forecasts with $\lambda = 0.97$. For a color version of this figure, see www.iste.co.uk/dimotikalis/analysis2.zip*

5.5. Conclusion

Observing the four series, it is clear that the decay factor has no effect on both errors (RMSE and MAPE). As λ varies, there are no changes on the two errors. We remember that, for the EWMA model, when λ increases, there is more weight on the recent values. Among the four values of λ used, we realize that $\lambda = 0.97$ is the best and for the three out-of-samples considered, we obtain good results on a wider out-of-sample period. This shows us that it is better to forecast exchange rate volatility considering a wider in-sample period also. We advise using 365.25 trading days based on our results and the fact that Forex markets operate even on weekends and holidays.

5.6. Acknowledgments

Jean-Paul Murara would like to thank the International Science Programme (ISP, Uppsala University) and the Wimas Group for the financial support, allowing this research paper to be written. Thanks also go to the Division of Applied Mathematics, School of Education, Culture and Communication, Mälardalen University, for creating an excellent environment for research in mathematics and applied mathematics.

5.7. References

Andersen, T.G., Bollerslev, T., Christoffersen, P.F., and Diebold, F.X. (2005). Volatility forecasting. Working Paper. National Bureau of Economic Research, Cambridge, MA.

Bollen, B. (2015). What should the value of lambda be in the exponentially weighted moving average volatility model? *Applied Economics*, 47(8), 853–860.

Hull, J. (2006). *Options, Futures, and Other Derivatives*. Pearson Prentice Hall, Englewood Cliffs.

J.P. Morgan (1996). RiskMetrics. Technical Document. J.P. Morgan/Reuters, New York.

National Bank of Rwanda (N/A). Exchange rate. [Online]. Available at: https://www.bnr.rw/footer/quick-links/exchange-rate/?txbnrcurrencyma-nagermaster%5Baction%5D=archive&txbnrcurrencymanagermaster%5Bco-ntroller%5D=Currency&cHash=9b3b8a3170a02e5876e4a1be17720fec [Accessed 3 January 2019].

Winters, P.R. (1960). Forecasting sales by exponentially weighted moving averages. *Management Science*, 6(3), 324–342.

6

An Arbitrage-free Large Market Model for Forward Spread Curves

Before the financial crisis started in 2007, the forward rate agreement contracts could be perfectly replicated by overnight indexed swap zero coupon bonds. After the crisis, the simply compounded, risk-free, overnight indexed swap forward rate became less than the forward rate agreement. Using an approach proposed by Cuchiero, Klein and Teichmann, we construct an arbitrage-free market model, where the forward spread curves for a given finite tenor structure are described as a mild solution to a boundary value problem for a system of infinite-dimensional stochastic differential equations. The constructed financial market is large: it contains infinitely many overnight indexed swap zero coupon bonds and forward rate agreement contracts, with all possible maturities. We also investigate the necessary assumptions and conditions which guarantee existence, uniqueness and non-negativity of solutions to the obtained boundary value problem.

6.1. Introduction and background

In the last decades of the previous century and for the duration of the third millennium, financial derivatives have significantly affected finance and the global market. In terms of underlying assets, the derivative market is massive, meaning that it is much larger than the stock market and, in terms of value, it is several times the world gross domestic product. In addition, derivative markets have been considered as the core of the financial crisis that began in 2007. That is, many derivative products which were constructed from a portfolio of risky mortgages became worthless after house prices decreased in the United States. Since then, many banks and financial institutions have changed their proxies in considering the term "risk-free" interest rate (Hull

---

Chapter written by Hossein NOHROUZIAN, Ying NI and Anatoliy MALYARENKO.

2015). Indubitably, the importance and impact of proper and accurate studies in this vast and significant field of mathematics is vital. Therefore, we attempt to develop an algebraic method for pricing financial contracts in the post-crisis financial market. This will also include calculating the forward rates. In this chapter, our objectives will be to:

– review the theories of constructing a large financial market model;

– construct an equivalent separating measure and prove its uniqueness;

– prove the existence, uniqueness and non-negativity of solutions to the system of SDEs describing the dynamics of a constructed large market.

We should emphasize that a large financial market model can include infinitely many assets (or more specifically, bonds). Furthermore, we will focus on the Heath–Jarrow–Morton (HJM) framework that describes no-arbitrage conditions that must be satisfied by a model of yield curves. Let us start with some preliminaries.

6.1.1. *Term-structure (interest rate) models*

To begin with, in derivative security models, the underlying assets are securities, whereas in term-structure models, the underlying assets are interest rates. In term structure models, the current value/price of a *default-free* (*risk-free*) *discount bond* for different maturities is called the *term-structure* of interest rate (Kijima 2013). Furthermore, interest rate derivatives are designed to protect an investor from huge losses against dramatic changes in interest rates. Bond options, swap options (swaptions), cap options (portfolios of caplets) and floor options (portfolios of floorlets) are important interest rate derivatives and are used to secure borrowing/lending from huge losses against dramatic changes in interest rates. Several different interest rate models have been developed to price interest rate derivatives. First, we denote the money market account by $B(t)$, the market price of a default-free discount bond by $P(t,T)$, the instantaneous interest rate (spot rate) by $r(t)$ and the instantaneous forward rate by $f(t,T)$, where

$$\text{for} \quad t \leq T, \quad r(t) = -\frac{\partial}{\partial T} \ln P(t,T)\big|_{T=t}, \quad f(t,T) = -\frac{\partial}{\partial T} \ln P(t,T).$$

On the contrary, we have the following relations between bonds and forward rates for both stochastic spot rate and deterministic spot rate

$$P(t,T) = \exp\left(-\int_t^T f(t,s)ds\right), \quad P(t,T) = \exp\left(-\int_t^T r(s)ds\right) = \frac{B(t)}{B(T)}.$$

Now, following Hull (2015) and Kijima (2013), we mention and categorize some of the most noteworthy and frequently used interest rate models in the following groups:

a) *Spot rate models:*

i) *Equilibrium models:* Rendleman–Bartter (Rendleman and Bartter 1980), Vasicek (Vasicek 1977), Cox–Ingersoll–Ross (CIR) (Cox *et al.* 1985) and Longstaff–Schwartz stochastic volatility model (Longstaff and Schwartz 1992).

ii) *No-arbitrage models:* Ho–Lee (Ho and Lee 1986), one-factor Hull–White (Hull and White 1990), Black–Derman–Toy (Black *et al.* 1990), Black–Karasinski (Black and Karasinski 1991) and two-factor Hull–White(Hull and White 1994).

b) *Forward rate models:*

i) Black (1976), discrete and continuous HJM (Heath *et al.* 1990, 1992) and LIBOR Market Model (LMM/BGM) (Brace *et al.* 1997).

6.1.2. *Forward-rate models versus spot-rate models*

The models mentioned above have different characteristics and might be useful for specific applications. For example, in the Vasicek model, the spot rate can become negative with positive probability. The Black model is easy to work for plain vanilla (relatively simple derivative) options, whereas HJM and LMM are more suitable for working with exotic (relatively complicated derivative) options. However, a problem with equilibrium models is that they do not fit today's term structure of interest rates (i.e. the term structure of interest rates is an output), whereas no-arbitrage models are designed to be consistent with today's term structure of interest rates (the term structure of interests rates is an input). The spot rate models in general have two important limitations. First, they usually involve only one factor or source of uncertainty. Second, they are not capable of choosing the volatility structure. The HJM and LMM, however, can be used to involve several factors and sources of uncertainty (Hull 2015). The HJM and LMM models also allow us to specify more realistic volatility structures to construct an interest rate model. That is, these models can be used in the evaluation of two (or more) yield curves. These curves can, for example, be LIBOR zero curves and *overnight indexed swap* (OIS) curves. For our purposes, we will focus on the HJM framework, which we briefly describe in the following section.

6.1.3. *The Heath–Jarrow–Morton framework*

In 1990 and 1992, David Heath, Bob Jarrow and Andy Morton (HJM) introduced a new framework in interest rate models (Heath *et al.* 1990, 1992). The defined framework describes no-arbitrage conditions which must be fulfilled by a yield curve model for some ultimate maturity τ (usually 20 or 30 years hence). The HJM model explains the dynamics of the forward rate curve $\{f(t,T,\tau), \quad 0 \leq t \leq T \leq \tau\}$.

In the HJM framework, the evolution of the forward curve satisfies the following stochastic differential equation (SDE) (Glasserman 2004):

$$df(t,T) = \mu(t,T)dt + \boldsymbol{\sigma}^\top(t,T)d\mathbf{W}(t). \quad [6.1]$$

where $\mathbf{W}$ is the standard d-dimensional Brownian motion and d represents the number of factors (sources of uncertainty). $\mu(t,T)$ is the drift structure and $\boldsymbol{\sigma}^\top(t,T)$ is the volatility structure. Moreover, the drift and volatility structures are $\mathbb{R}^d$-valued and can be either stochastic, or can depend on the current and past level of the forward rate.

Risk-neutral evaluation: In the HJM framework, the arbitrage-free dynamics of the forward curve, using risk-neutral evaluation, satisfies the following SDE (Glasserman 2004):

$$df(t,T) = \left(\boldsymbol{\sigma}^\top(t,T)\int_t^T \boldsymbol{\sigma}(t,u)du\right)dt + \boldsymbol{\sigma}^\top(t,T)d\mathbf{W}(t).$$

6.1.4. *Construction of our model*

Before the financial crisis started in 2007, LIBOR rates were commonly used as risk-free rates, whereas in the post-crisis market, OIS rates are considered the new proxies for risk-free rates. Moreover, an OIS is a swap contract to exchange cash flows at a fixed rate (called the OIS rate) for the geometric average of the overnight rates during the same period. For a fixed period (e.g. three months), the OIS rates are generally lower than LIBOR rates which yields to the so-called LIBOR-OIS spread (Hull 2015).

In order to price financial instruments under collateral and *forward rate agreement* (FRA) rates, we follow the approach considered in Filipović and Trolle (2013) and Cuchiero *et al.* (2016a). That is, *OIS zero coupon bonds* are considered to be the basic traded instruments, and they play the default-free zero coupon bonds' role in the old setting. Thus, $B(t) = \exp\left(\int_0^t r(s)ds\right)$ is now the OIS (risk-free) bank account with r representing the OIS short rate.

Now, we will follow the HJM framework and modify equation [6.1] such that we will have a set of initial forward spread curves as well as a set of forward spread curves with different maturities. The set of forward spread curves will, of course, have different sources of uncertainty. In other words, we construct an arbitrage-free market model, where the forward spread curves for a given finite tenor structure are described as a mild solution to a *boundary value problem* (BVP) for a system of infinite-dimensional stochastic differential equations. The constructed financial market is *large*: it contains infinitely many OIS zero coupon bonds and FRA contracts with all possible maturities.

In summary, we consider and pursue the following concepts and objectives in this chapter. In section 6.2, we will go through the definitions of small and large financial markets, no asymptotic free lunch conditions in a large market and present the *fundamental theorem of asset pricing* (FTAP) for a large market. Also, we construct a unique risk-neutral probability measure (equivalent martingale probability measure). Then, in section 6.3, we construct a system of stochastic partial differential equations and discuss the necessary conditions and assumptions which guarantee existence, uniqueness and non-negativity of solutions to the obtained BVP. Finally, we will close this chapter with a conclusion–future works section.

6.2. Construction of a market with infinitely many assets

Some of the important studies in the theory of market models with infinitely many assets have been conducted by Björk *et al.* (1997), De Donno and Pratelli (2005), and Ekeland and Taflin (2005). However, using such a theory, according to Taflin (2011), might lead to some difficulties. That is, the construction of such a market does not imply that the market is complete. In other words, a market without any arbitrage opportunities is complete if and only if there exists a unique *risk-neutral probability measure* (RNPM) (Kijima 2013).

Some other approaches were developed to overcome these sorts of difficulties. One of them is an approach considered by Klein *et al.* (2016). In this approach, we can see a system of equations (for different tenors), and for each equation, there exists a RNPM. However, to our knowledge, we cannot prove if the considered RNPMs in different equations are the same. Mathematically speaking, the ith component of the market satisfies *no asymptotic free lunch condition* (NAFLC) if and only if there exists an *equivalent martingale probability measure* (EMPM), say Q_i. However, we cannot prove if such Q_is are the same or not. Another approach is that found in Cuchiero *et al.* (2016b) which we will review in more detail in the following section.

6.2.1. *The Cuchiero–Klein–Teichmann approach*

First, following Cuchiero *et al.* (2016b), a finite market is referred to as a *small* market and an infinite market as a *large* market. A large market can be seen as the limit of a sequence of small markets. At a glance, the objective is to prove a version of the FTAP for the notion of *no asymptotic free lunch with vanishing risk* (NAFLVR). We begin with summarizing some important concepts and notations. That is:

1) $S_n(t)_{t\in[0,1]}$ are defined on a filtered probability space $(\Omega, \mathfrak{F}, (\mathfrak{F}_t)_{t\in[0,1]}, \mathsf{P})$.

2) $\{S_n(t)\colon n \geq 0,\ 0 \leq t \leq 1\}$ is a sequence of semimartingales with $S_0(t) = 1$.

3) $\{S_i(t)\colon 0 \leq i < n,\ 0 \leq t \leq 1\}$ is a subset defining a *finite market model* n.

4) A *large financial market* corresponds to the limit of a sequence of finite market models n, generated by the sequence $\{S_n(t)\colon n \geq 0,\ 0 \leq t \leq 1\}$.

5) $C = (K_0 - L^0_{\geq 0}) \cap L^\infty$ denotes a convex cone of bounded claims at price 0, where:

– K_0 is a set of admissible generalized portfolios' terminal values at $t = 1$;

– L^0 is the set of all measurable functions;

– $L^0_{\geq 0}$ is the set of all non-negative measurable functions;

– L^∞ is the space of bounded functions;

– $L^\infty_{\geq 0}$ is the space of non-negative bounded functions.

6) *No asymptotic free lunch with vanishing risk* is defined by $\overline{C} \cap L^\infty_{\geq 0} = \{0\}$.

7) $\mathscr{S}$ denotes the set of all semimartingales defined on $[0,1]$, starting from 0, and adapted to the filtration $\{\mathfrak{F}_t : 0 \leq t \leq 1\}$.

8) $\mathscr{K}$ denotes the set of simple predictable strategies.

9) I denotes a parameter space.

10) $0 = \tau_0 \leq \tau_1 \leq \cdots \leq \tau_{\ell+1} = 1$ are stopping times.

11) ς denotes the ς-admissible generalized portfolio strategies.

REMARK 6.1.– *NAFLVR is equivalent to the following conditions:*

i) no unbounded profit with bounded risk (NUPBR);

ii) no arbitrage for large markets (NA).

Now, we follow the CKT approach. Émery (1979) defined the metric on $\mathscr{S}$ by

$$d_{\mathscr{S}}(X_1, X_2) = \sup \mathsf{E}[\min\{\sup_{t \in [0,1]} |(\mathscr{K} \cdot (X_1 - X_2))_t|, 1\}], \qquad [6.2]$$

where the outer supremum is taken over the set of all bounded predictable processes $\mathscr{K}$ bounded by 1 in absolute value (not only over all simple predictable processes). Mémin (1980) proved that taking supremum over the set of *simple* predictable strategies bounded by 1 in absolute value yields

$$K(t) = \sum_{i=0}^{\ell} k_i \mathbb{1}_{(\tau_i, \tau_{i+1}]}(t), \quad \text{for} \quad i = 0, \ldots, \ell, \quad k_i \in \mathscr{K},$$

where ℓ is the positive integer, and k_i is the $\mathfrak{F}_{\tau_i}$-measurable random variable. The obtained metric is equivalent to the metric [6.2].

Now, for (parameter space) $I \subseteq [0, \infty)$ and each positive integer n, we define a family $\mathscr{A}^n$ of subsets of I satisfying the following conditions:

i) each set in $\mathscr{A}^n$ contains exactly n elements;

ii) if $A^1, A^2 \in \bigcup_{n=1}^\infty \mathscr{A}^n$, then $A^1 \cup A^2 \in \bigcup_{n=1}^\infty \mathscr{A}^n$.

Here, we can present and formulate the mathematical definitions of small and large financial markets.

DEFINITION 6.1 (Small financial market).– *A small financial market indexed by a set* $A \in \bigcup_{n=1}^{\infty} \mathscr{A}^n$ *is a set* $\mathscr{X}_1^A \subset \mathscr{S}$ *which satisfies the following conditions:*

i) Bounded: each element of $\mathscr{X}_1^A$ *is bounded from below by* -1.

ii) Monotonicity: if $A^1, A^2 \in \bigcup_{n=1}^{\infty} \mathscr{A}^n$ *with* $A^1 \subset A^2$, *then* $\mathscr{X}_1^{A^1} \subset \mathscr{X}_1^{A^2}$.

iii) Concatenation property: if G and H are the bounded predictable processes with $G \geq 0, H \geq 0, GH = 0$, *then for all* $X, Y \in \mathscr{X}_1^A$ *such that*

$$Z = G \cdot X + H \cdot Y \geq -1,$$

we have $Z \in \mathscr{X}_1^A$.

The set $\mathscr{X}_1^A$ is called *the set of one-admissible portfolio wealth processes in the small financial market A*. Moreover, the *set of all one-admissible portfolio wealth processes with respect to strategies that include at most* n *assets* is given by

$$\mathscr{X}_1^n = \bigcup_{A \in \mathscr{A}^n} \mathscr{X}_1^A.$$

DEFINITION 6.2 (Components in a large financial market).–

1) By $\overline{(\cdot)}^{\mathscr{S}}$, *we mean the closure in the Émery topology. The set of all one-admissible generalized portfolio wealth processes in the large financial market is then*

$$\mathscr{X}_1 = \overline{\bigcup_{n=1}^{\infty} \mathscr{X}_1^n}^{\mathscr{S}}.$$

2) $\mathscr{X}$ *denotes the set of all admissible generalized portfolio wealth processes in the large financial market and is given by*

$$\mathscr{X} = \bigcup_{\varsigma > 0} \varsigma \mathscr{X}_1.$$

3) The evaluations of elements of $\mathscr{X}_1$ *and* $\mathscr{X}_2$ *at time terminal* $T = 1$ *are*

$$K_0 = \{X(1) \colon X \in \mathscr{X}\}, \qquad K_0^1 = \{X(1) \colon X \in \mathscr{X}_1\}.$$

DEFINITION 6.3 (NAFLVR (Cuchiero *et al.* 2016b)).– *Let C be the set of all bounded random variables in a convex cone* C_0 *defined by*

$$C_0 := K_0 - L_{\geq 0}^{\infty}, \quad C := C_0 - L^{\infty},$$

where the minus operation here means $C_0 = \{Y \colon Y \leq X \text{ for some } X \in K_0\}$. *Also, let* $\overline{C}$ *be the closure of C in* L^{∞}. *Then, by definition, the set* $\mathscr{X}$ *satisfies no asymptotic free lunch with vanishing risk (NFLVR) if*

$$\overline{C} \cap L_{\geq 0}^{\infty} = \{0\}.$$

DEFINITION 6.4 (Equivalent separating measure (ESM) (Cuchiero *et al.* 2016b)).– *The set $\mathscr{X}$ satisfies the equivalent separating measure property if there exists a measure* Q *equivalent to* P, *i.e.* Q $\sim$ P, *such that*

$$\mathsf{E}^{\mathsf{Q}}[X(1)] \leq 0, \qquad X \in \mathscr{X}.$$

THEOREM 6.1 (FTAP for a large market (Cuchiero *et al.* 2016b)).– *The set $\mathscr{X}$ satisfies no asymptotic free lunch with vanishing risk if and only if it satisfies the equivalent separating measure property.*

6.2.2. *Adapting Cuchiero–Klein–Teichmann's results to our objective*

An obstacle for us is that in the CKT's result $I \subseteq [0,\infty)$. In our case, we have m numbers of tenors. Thus, our desirable parameter space has the following form:

$$I = [0,\infty)^m, \quad m \in \mathbb{N}^+. \qquad [6.3]$$

Therefore, we would like to prove that Theorem 6.1 remains true when the set I (parameter space) has the form [6.3].

THEOREM 6.2.– *For m-dimensional parameter space, i.e. $I = [0,\infty)^m$, Theorem 6.1 (CKT's FTAP for a large market) remains true.*

PROOF.– Call a property of the *one-dimensional* parameter space $[0,\infty)$ *special* if there is a positive integer m such that the set $[0,\infty)^m$ *does not have* this property. For example, the space $[0,\infty)$ is ordered by the relation $\leq$, while the spaces $[0,\infty)^m$ are not ordered.

Careful analysis of Theorem 3.1, and its proof in Cuchiero *et al.* (2016b, Section 7) shows that no special properties of the one-dimensional parameter space $[0,\infty)$ have been used. This fact completes the proof. □

6.3. Existence, uniqueness and non-negativity

In this section, we investigate the necessary conditions for existence, uniqueness and non-negativity of solutions to some desirable stochastic equations within the HJM framework. A forward spread curve in our setting is a solution to some *stochastic partial differential equations* (SPDE) with some initial conditions which describes the dynamic of some price processes in a large financial market (with infinitely many assets/bonds).

6.3.1. *Existence and uniqueness: mild solutions*

To begin with, we introduce the following notations:

1) $P(t,T)$ denotes the OIS zero coupon bond price.

2) $\delta = \{\delta_i,\ 1 \le i \le m,\ m \in \mathbb{N}^+\}$ are the tenor dates.

3) $L_t(T,T+\delta)$ is the (normalized) FRA rate and $L_T(T,T+\delta)$ are the LIBOR rates.

4) $L_t^{\text{OIS}}(T,T+\delta)$ is the (normalized) simply compounded OIS forward rate.

5) $S^\delta(t,T) = \dfrac{1+\delta L_t(T,T+\delta)}{1+\delta L_t^{\text{OIS}}(T,T+\delta)}$ is called the *multiplicative spread.*

6) $\eta_t^i(T) = \partial_T \ln S^{\delta_i}(t,T), 1 \le i \le m$ are the *instantaneous forward spread curves* and together with $\eta_t^0(T) = \partial_T \ln P(t,T)$ form a $\mathbb{R}^{m+1}$-valued *random field*

$$\{\boldsymbol{\eta}_t(T) \colon 0 \le t \le T < \infty\}.$$

7) $\boldsymbol{\theta}_t(s) = \boldsymbol{\eta}_t(t+s)$ is called the *Musiela parametrization* after Musiela (1993), where $s \in [0,\infty)$.

8) H_k^λ is the *Filipović space* after Filipović (2001), i.e. the Hilbert space of absolutely continuous functions $\mathbf{h} \colon [0,\infty) \to \mathbb{R}^k$ satisfying the condition

$$\|\mathbf{h}\|^2 = \|\mathbf{h}(0)\|_k^2 + \int_0^\infty \|\mathbf{h}'(s)\|_k^2 e^{\lambda s}\,\mathrm{d}s < \infty,$$

where $\|\cdot\|$ denotes the norm in $\mathbb{R}^k$ with $k \in \{1,d,m+1\}$ and $\lambda > 0$.

9) $\boldsymbol{W}$ is the d-dimensional Wiener process, $\theta_t^i \in H_1^\lambda$ and $\boldsymbol{\theta}_t \in H_{m+1}^\lambda$.

10) $\kappa^i \colon H_{m+1}^\lambda \to H_1^\lambda$ and $\kappa^i(\boldsymbol{\theta}_t)$ are the drift coefficients/functions.

11) $\zeta^i \colon H_{m+1}^\lambda \to L(\mathbb{R}^d, H_1^\lambda)$, where $L(\cdot,\cdot)$ is the space of linear operators and $\zeta^i(\boldsymbol{\theta}_t)$ are the diffusion coefficients.

12) $(S_s)_{\ge 0}$ is the shift semi-group on H_{m+1}^λ, i.e. $S_s\boldsymbol{h} = \boldsymbol{h}(s+\cdot)$.

13) $\dfrac{d}{ds}$ becomes the infinitesimal operator (generator) of the strongly continuous semi-group of shifts $(S_s)_{\ge 0}$.

Following Cuchiero *et al.* (2016a), we consider the technicality presented in Heath *et al.* (1992), Da Prato and Zabczyk (2014) and Cuchiero *et al.* (2016b), and after doing a little algebra (see Malyarenko *et al.* (2020)), the Musiela parametrization can be expressed by the following system of stochastic integral equations

$$\theta_t^i = S_t\eta_0^i + \int_0^t S_{t-s}\kappa^i(\boldsymbol{\theta}_s)\,ds + \int_0^t S_{t-s}\zeta^i(\boldsymbol{\theta}_s)\,d\boldsymbol{W}_s, \qquad 0 \le i \le m. \qquad [6.4]$$

According to Da Prato and Zabczyk (2014), a H^{λ}_{m+1}-valued process $\boldsymbol{\theta}_t$ satisfying the above system of equations is called a *mild solution* to the system of the following SPDEs:

$$\begin{cases} d\theta^i_t = \left(\dfrac{d}{ds}\theta^i_t + \kappa^i(\boldsymbol{\theta}_t)\right) dt + \zeta^i(\boldsymbol{\theta}_t)\, d\boldsymbol{W}_t, \\ \theta^i_0 = \eta^i_0. \end{cases} \tag{6.5}$$

Within the HJM framework, the drift coefficients, $\kappa^i(\boldsymbol{\theta}_t)$, can be uniquely determined by the rest of the data (see Cuchiero *et al*. (2016a, Equation (4.6))).

REMARK 6.2.– *In Cuchiero* et al. *(2016a),* $\zeta^i \colon H^{\lambda}_{m+1} \to H^{\lambda}_d$ *and in our work* $\zeta^i \colon H^{\lambda}_{m+1} \to L(\mathbb{R}^d, H^{\lambda}_1)$*. Since* $L(\mathbb{R}^d, H^{\lambda}_1)$ *and* H^{λ}_d *are isomorphic, both definitions are equivalent.*

REMARK 6.3.– *Equations* [6.4] *and* [6.5] *in Cuchiero* et al. *(2016a) and Filipović* et al. *(2010) have one additional term for considering dramatic changes (jumps) in the dynamic of* θ^i_t*, which we omit.*

Following Cuchiero *et al*. (2016a) (see also Filipović *et al*. (2010, Assumption 3.1)), we introduce the following assumptions to establish the existence and uniqueness of a mild solution to equation [6.5].

ASSUMPTION 6.1 (Existence and uniqueness (Cuchiero *et al*. 2016a)).– *The growth and Lipschitz continuity conditions on the volatility function* ζ^i *for all* $i = 1, \ldots, m$ *can be formulated by the following axioms:*

($\mathscr{E}$1) *For all* $i = 1, \ldots, m$*, we have* $\zeta^i \colon H^{\lambda}_{m+1} \to L(\mathbb{R}^d, H^{\lambda,0}_1)$*, where* $H^{\lambda,0}_k := \{\boldsymbol{h} \in H^{\lambda}_k : \lim_{s\to\infty} \|\boldsymbol{h}(s)\|_k = 0\}$*, for* $k \in \{1, d\}$*.*

($\mathscr{E}$2) *For all* $i = 1, \ldots, m$*, there exist positive constants* K_i, L_i, M_i *such that*

- ($\mathscr{E}$2.1) $\|\int_0^s \zeta^i(\boldsymbol{h})(u)du\|_d \leq K_i$*, for all* $\boldsymbol{h} \in H^{\lambda}_{m+1}$*,* $s \in [0, \infty)$*;*
- ($\mathscr{E}$2.2) $\|\zeta^i(\boldsymbol{h}_1) - \zeta^i(\boldsymbol{h}_2)\|_{\lambda,d} \leq L_i \|\boldsymbol{h}_1 - \boldsymbol{h}_2\|_{\lambda,m+1}$*, for all* $\boldsymbol{h}_1, \boldsymbol{h}_2 \in H^{\lambda}_{m+1}$*;*
- ($\mathscr{E}$2.3) $\|\zeta^i(\boldsymbol{h})\|_{\lambda,d} \leq M_i$ *for all* $\boldsymbol{h} \in H^{\lambda}_{m+1}$*.*

Now, following Cuchiero *et al*. (2016a, Proposition 4.4), we present the following proposition (see also Filipović *et al*. (2010, Assumption 4.11)).

PROPOSITION 6.1.– *If the axioms in Assumption 6.1 are met, then for every* $i \in \{0, \ldots, m\}$*,* $\kappa^i(H^{\lambda}_{m+1}) \subseteq H^{\lambda,0}_1$ *holds. Furthermore, for all* $\boldsymbol{h}_1, \boldsymbol{h}_2 \in H^{\lambda}_{m+1}$*, there exist constants* $Q_i > 0$ *such that*

$$\|\kappa^i(\boldsymbol{h}_1) - \kappa^i(\boldsymbol{h}_2)\|_{\lambda,1} \leq Q_i \|\boldsymbol{h}_1 - \boldsymbol{h}_2\|_{\lambda,m+1}.$$

Finally, the existence and uniqueness of the solution to a system of SPDE [6.5] is established by the following theorem.

THEOREM 6.3 (Existence and uniquness (Cuchiero *et al.* 2016a)).– *If the axioms in Assumption 6.1 are met, then for all $T \in \mathbb{R}^+$, and each initial $\theta_0^i \in H_{m+1}^\lambda$, there exists a unique adapted cÃ dlÃ g, mean-square continuous mild H_{m+1}^λ-valued solution $(\theta_t^i)_{t\geq 0}$ such that*

$$\mathsf{E}\left[\sup_{t\in[0,T]} \|\theta_t^i\|_{\lambda,m+1}^2\right] < \infty.$$

6.3.2. *Non-negativity of solutions*

In this section, we investigate the necessary and sufficient conditions that give non-negative forward spread curves, i.e. $\theta_t^i \geq 0$ for all $t \in [0,T]$, for given non-negative initial curves, i.e. $\theta_0^i \geq 0$.

According to Filipović (2001), positivity of the diffusion part in equation [6.5] can be seen in the result of Kotelenez (1992) and Milian (2002). Also, the non-negativity of the solution to equation [6.5] is proved by Nakayama (2004). Let us discuss Nakayama's approach and apply his result to our model in the following steps:

Step 1. (Notations) We introduce the following notations:

1) V is a separable Hilbert space with inner product $\langle\cdot,\cdot\rangle_V$ and norm $\|\cdot\|_V$.

2) U is a separable Hilbert space with inner product $\langle\cdot,\cdot\rangle_U$.

3) Q is a trace class strictly positive operator on U.

4) $U_0 = Q^{1/2}(U)$ is a separable Hilbert space, with the following inner product $\langle u,v\rangle_{U_0} = \langle Q^{-1/2}u, Q^{-1/2}v\rangle_U$, where $u,v \in U_0$ and norm $\|\cdot\|_{U_0}$.

5) $A\colon V \to V$ is a linear operator.

6) α_i, $i = 1,2,\ldots$ is an orthonormal basis of U_0.

7) β_j, $j = 1,2,\ldots$ is an orthonormal basis of V.

8) D and D^2 denote the first- and second-order Fréchet derivatives.

9) ρ and ϱ are the distance functions.

Step 2. Put $A = \frac{d}{ds}$ and rewrite equation [6.5] in vector form. That is,

$$\begin{cases} d\boldsymbol{\theta}_t = (A\boldsymbol{\theta}_t + \boldsymbol{\kappa}(\boldsymbol{\theta}_t))\,dt + \boldsymbol{\zeta}(\boldsymbol{\theta}_t)\,d\mathbf{W}_t, \\ \boldsymbol{\theta}_0 = \boldsymbol{\eta}_0, \end{cases} \qquad [6.6]$$

where $\boldsymbol{\theta}_t \in H_{m+1}^\lambda$, $\boldsymbol{\kappa}\colon H_{m+1}^\lambda \to H_{m+1}^\lambda$ and $\boldsymbol{\zeta}\colon H_{m+1}^\lambda \to L(\mathbb{R}^d, H_{m+1}^\lambda)$.

Step 3. (Lipschitz-continuity and Hilbert–Schmidt operators) Nakayama (2004) considers a class of equations that includes equation [6.6] as a particular case. In our notation, Nakayama (2004) requires $\boldsymbol{\zeta}$ to map H^{λ}_{m+1} to the linear space of Hilbert–Schmidt operators from U_0 to H^{λ}_{m+1}. In our case, $U = \mathbb{R}^d$, Q is the identity operator on U. Then, $U_0 = U$. For any linear operator from the finite-dimensional space U_0 to H^{λ}_{m+1}, the sum of squares of its eigenvalues is obviously finite, and this operator is Hilbert–Schmidt.

Now, we can claim that the conditions for Lipschitz-continuous bounded mapping in Nakayama (2004) are equivalent to the conditions in Assumption 6.1 and Proposition 6.1.

ASSUMPTION 6.2 (Non-negativity (Nakayama 2004)).–
($\mathscr{N}$1) Assume $\zeta^j(\boldsymbol{\theta}) = \boldsymbol{\zeta}(\boldsymbol{\theta})\beta_j$, $j = 1,2,\ldots$, $\boldsymbol{\theta} \in H^{\lambda}_{m+1}$ *are twice Fréchet differentiable and bounded. That is,*

$$\sup\{\|D\zeta^j(\boldsymbol{\theta})(\boldsymbol{h})\| : \boldsymbol{h} \in H^{\lambda}_{m+1},\ \|\boldsymbol{h}\| \leq 1,\ \boldsymbol{\theta} \in H^{\lambda}_{m+1}\},$$

$$\sup\{\|D^2\zeta^j(\boldsymbol{\theta})(\boldsymbol{h}_1,\boldsymbol{h}_2)\| : \boldsymbol{h}_1,\boldsymbol{h}_2 \in H^{\lambda}_{m+1},\ \|\boldsymbol{h}_1\| \leq 1,\ \|\boldsymbol{h}_2\| \leq 1,\ \boldsymbol{\theta} \in H^{\lambda}_{m+1}\}.$$

($\mathscr{N}$2) Let $n \in \mathbb{N}^+$, $\rho_n \colon H^{\lambda}_{m+1} \to H^{\lambda}_{m+1}$ *and* $\rho_n(\boldsymbol{\theta}) = 1/2\sum_{j=1}^{n} D\zeta^j(\boldsymbol{\theta})\zeta^j(\boldsymbol{\theta})$ *for* $\boldsymbol{\theta} \in H^{\lambda}_{m+1}$. *We assume that there exists a map* $\rho_n \colon H^{\lambda}_{m+1} \to H^{\lambda}_{m+1}$, *such that for all* $\boldsymbol{\theta} \in H^{\lambda}_{m+1}$, *the following holds:*

$$\lim_{n\to\infty} \|\rho_n(\boldsymbol{\theta}) - \rho(\boldsymbol{\theta})\| = 0.$$

($\mathscr{N}$3) Assume for any $\boldsymbol{h}_1,\boldsymbol{h}_2 \in H^{\lambda}_{m+1}$ *and all* $n \in \mathbb{N}^+$, *there exists a constant M such that*

$$\|\rho_n(\boldsymbol{h}_1) - \rho_n(\boldsymbol{h}_2)\| \leq M\|\boldsymbol{h}_1 - \boldsymbol{h}_2\|.$$

Now, for any $\boldsymbol{\eta}_0 \in H^{\lambda}_{m+1}$, let $(\boldsymbol{\theta}_t)_{t\in[0,T]}$ be the mild solution to the SDE [6.6], or equivalently $(\theta^i_t)_{t\in[0,T]}$ satisfies equation [6.4].

Step 4. Define $\mathscr{C}^1$ as the set of all functions $g \colon [0,T] \to \mathbb{R}^d$ such that g is continuous and piecewise continuously differentiable with $g(0) = \mathbf{0}$. For any $g \in \mathscr{C}^1$, let $\boldsymbol{\xi}(t;\boldsymbol{\eta}_0,g)$ be the unique solution of the following *deterministic* differential equation:

$$\begin{cases} d\boldsymbol{\xi}_t = A\boldsymbol{\xi}_t + (\boldsymbol{\kappa} - \rho)(\boldsymbol{\xi}(t;\boldsymbol{\eta}_0,g)) + \boldsymbol{\zeta}(\boldsymbol{\xi}(t;\boldsymbol{\eta}_0,g))\,dg_t, \\ \boldsymbol{\xi}(0;\boldsymbol{\eta}_0,g) = \boldsymbol{\eta}_0. \end{cases} \qquad [6.7]$$

Equivalently, denote $\xi^i(t;\cdot) = \xi^i(t;\eta^i_0,g)$, then for $t \in [0,T]$, we have

$$\xi^i(t;\eta^i_0,g) = S_t\eta^i_0 + \int_0^t S_{t-s}(\kappa^i - \rho)(\xi^i(s;\cdot))ds + \int_0^t S_{t-s}\zeta^i(\xi^i(s;\cdot))dg(s)ds.$$

Step 5. Let us rewrite $V = \{\sum_{j=1}^{\infty} b_j \boldsymbol{\beta}_j : \sum_{j=1}^{\infty} b_j^2 < \infty\}$. Furthermore, let us define $Z = \{\boldsymbol{\theta}_t \in H_{m+1}^{\lambda} : \theta_t^i \geq 0,\ t \geq 0,\ 0 \leq i \leq m\}$. Now, we are able to see the non-negativity of solutions to our SPDEs in the following proposition (see Proposition 1.1 in Nakayama (2004) for details and proof, as well as Proposition 4.19 in Filipović *et al*. (2010)).

PROPOSITION 6.2.– *The following three conditions are equivalent:*

(a) For every $\eta_0^i \in Z$ *(and* $\theta_0^i = \eta_0^i$*),* $\mathsf{P}\{\ \theta_t^i \in Z,\ \textit{for all}\ \ t \in [0,T]\ \} = 1$.

(b) For every $g \in \mathscr{C}^1$, $\eta_0^i \in Z$ *and* $t \in [0,T]$, *we have* $\xi(t;g) \in Z$.

(c) For every $\eta_0^i \in Z$ *and* $u \in \mathbb{R}^d$ *(the so-called "semigroup Nagumo's condition")*

$$\lim_{t \downarrow 0} \frac{1}{t}\, \varrho\left(S_t \boldsymbol{\eta}_0 + t\big(\boldsymbol{\kappa}(\boldsymbol{\eta}_0) - \rho(\boldsymbol{\eta}_0) + \boldsymbol{\zeta}(\boldsymbol{\eta}_0)u\big),\ Z\right) = 0,$$

where $\varrho(z,Z) = \inf\{\|z - \hat{z}\|_{H_{m+1}^{\lambda}} : \hat{z} \in Z\}$ *for any* $z \in H_{m+1}^{\lambda}$.

According to Nakayama (2004), Z is said to be invariant for SDE [6.6] if condition (a) is satisfied and is invariant for DE [6.7] if condition (b) is satisfied. The proposition above tells us Z is invariant for SDE [6.6] if and only if Z is invariant for DE [6.7]. Also, conditions (a) and (b) are equivalent to condition (c). As a result, we can conclude that the operator $\boldsymbol{\zeta} : H_{m+1}^{\lambda} \to L(\mathbb{R}^d, H_{m+1}^{\lambda})$ is indeed a Hilbert–Schmidt operator and the solution to the SPDE [6.5] exists and is unique and non-negative.

6.4. Conclusion and future works

In this chapter, we reviewed the construction of a large financial market model which can include infinitely many assets. Then, we constructed an equivalent separating measure in the defined large financial market model. After that, we showed that such a measure is unique for a finite dimensional parameter space and tenor structure. In this procedure, we also proved that the fundamental theorem of asset pricing for the large financial market model with a finite dimensional parameter space and tenor structure remains true. In addition, we have presented the necessary assumptions and conditions which guarantee the existence, uniqueness and non-negativity of the solution to our boundary value problem that describes the dynamics of forward rates in our defined large financial market model. Finally, we proved the existence, uniqueness and non-negativity of solutions to the described system.

In our future works, we will attempt to approximate the value of a contingent claim with underlying stochastic process $\boldsymbol{\theta}_t$. More specifically, we intend to develop an approximation method, using the so-called cubature on Wiener space (see Bayer and Teichmann (2008)), for the quantity $\mathsf{E}[f(\boldsymbol{\theta}_t)]$, where $f : H_{m+1}^{\lambda} \to \mathbb{R}$. This will be presented in forthcoming work.

6.5. References

Bayer, C. and Teichmann, J. (2008). Cubature on Wiener space in infinite dimension. *Proc. R. Soc. Lond. Ser. A Math. Phys. Eng. Sci.*, 464(2097), 2493–2516.

Björk, T., Di Masi, G., Kabanov, Y., Runggaldier, W. (1997). Towards a general theory of bond markets. *Finance Stoch.*, 1(1), 141–174.

Black, F. (1976). The pricing of commodity contracts. *J. Financ. Econom.*, 3(1), 167–179.

Black, F. and Karasinski, P. (1991). Bond and option pricing when short rates are lognormal. *Financial Analysts J.*, 47(4), 52–59.

Black, F., Derman, E., Toy, W. (1990). A one-factor model of interest rates and its application to treasury bond options. *Financial Analysts J.*, 46(1), 33–39.

Brace, A., Gatarek, D., Musiela, M. (1997). The market model of interest rate dynamics. *Math. Finance*, 7(2), 127–155.

Cox, J.C., Ingersoll, J.E., Ross, S.A. (1985). A theory of the term structure of interest rates. *Econometrica*, 53(2), 385–407.

Cuchiero, C., Fontana, C., Gnoatto, A. (2016a). A general HJM framework for multiple yield curve modeling. *Finance Stoch.*, 20(2), 267–320.

Cuchiero, C., Klein, I., Teichmann, J. (2016b). A new perspective on the fundamental theorem of asset pricing for large financial markets. *Theory Probab. Appl.*, 60(4), 561–579.

Da Prato, G. and Zabczyk, J. (2014). *Stochastic Equations in Infinite Dimensions*. Cambridge University Press, Cambridge.

De Donno, M. and Pratelli, M. (2005). A theory of stochastic integration for bond markets. *Ann. Appl. Probab.*, 15(4), 773–791.

Ekeland, I. and Taflin, E. (2005). A theory of bond portfolios. *Ann. Appl. Probab.* 15(2), 1260–1305.

Émery, M. (1979). Une topologie sur l'espace des semimartingales. In *Séminaire de Probabilités, XIII (Univ. Strasbourg, Strasbourg, 1977/78)*, vol. 721 of *Lect. Notes Math.*, 260–280. Springer, Berlin.

Filipović, D. (2001). *Consistency Problems for Heath–Jarrow–Morton Interest*, vol. 60 of *Lect. Notes Math.*, Springer-Verlag, Berlin.

Filipović, D. and Trolle, A.B. (2013). The term structure of interbank risk. *J. Financ. Econ.*, 109(3), 707–773.

Filipović, D., Tappe, S., Teichmann, J. (2010). Term structure models driven by Wiener processes and Poisson measures: Existence and positivity. *SIAM J. Financ. Math.*, 1(1), 523–554.

Glasserman, P. (2004). *Monte Carlo Methods in Financial Engineering*. Springer, New York.

Heath, D., Jarrow, R., Morton, A. (1990). Bond pricing and the term structure of interest rates: A discrete time approximation. *J. Financ. Quantitative Analysis*, 25(4), 419–440.

Heath, D., Jarrow, R., Morton, A. (1992). Bond pricing and the term structure of interest rates: A new methodology for contingent claims valuation. *Econometrica*, 60(1), 77–105.

Ho, T.S. and Lee, S.B. (1986). Term structure movements and pricing interest rate contingent claims. *J. Finance*, 41(5), 1011–1029.

Hull, J.C. (2015). *Options, Futures, and Other Derivatives*. Pearson Prentice Hall, New Jersey.

Hull, J.C. and White, A. (1990). Pricing interest-rate-derivative securities. *Rev. Financ. Studies*, 3(4), 573–592.

Hull, J.C. and White, A. (1994). Numerical procedures for implementing term structure models II: Two-factor models. *J. Derivatives*, 2(2), 37–48.

Kijima, M. (2013). *Stochastic Processes with Application to Finance*. Chapman & Hall/CRC, Florida.

Klein, I., Schmidt, T., Teichmann, J. (2016). No arbitrage theory for bond markets. In *Advanced Modelling in Mathematical Finance*, Kallsen, J. and Papapantoleon, A. (eds), vol. 189 of *Springer Proc. Math. & Statist.*, pp. 381–421, Springer, Berlin.

Kotelenez, P. (1992). Comparison methods for a class of function valued stochastic partial differential equations. *Probab. Theory Related Fields*, 93(1), 1–19.

Longstaff, F.A. and Schwartz, E.S. (1992). Interest rate volatility and the term structure: A two-factor general equilibrium model. *J. Finance*, 47(4), 1259–1282.

Malyarenko, A., Nohrouzian, H., Silvestrov, S. (2020). An algebraic method for pricing financial instruments on post-crisis market. In *Algebraic Structures and Applications*, Silvestrov, S., Malyarenko, A., Rančić, M. (eds). Springer Nature, Berlin.

Mémin, J. (1980). Espaces de semi martingales et changement de probabilité. *Z. Wahrsch. Verw. Gebiete*, 52(1), 9–39.

Milian, A. (2002). Comparison theorems for stochastic evolution equations. *Stochastics Stoch. Reports*, 72(1–2), 79–108.

Musiela, M. (1993). Stochastic PDEs and term structure models. *Journées internationales de finance, IGR–AFFI*, La Baule, France.

Nakayama, T. (2004). Viability theorem for SPDE's including HJM framework. *J. Math. Sci. Univ. Tokyo*, 11(1), 313–324.

Oertel, F. and Owen, M. (2007). On utility-based super-replication prices of contingent claims with unbounded payoffs. *J. Appl. Probab.*, 44(4), 880–888.

Rendleman, R.J. and Bartter, B. (1980). The pricing of options on debt securities. *J. Financ. Quantitative Analysis*, 15(1), 11–24.

Taflin, E. (2011). Generalized integrands and bond portfolios: Pitfalls and counter examples. *Ann. Appl. Probab.*, 21(1), 266–282.

Vasicek, O. (1977). An equilibrium characterization of the term structure. *J. Financ. Econ.*, 5(2), 177–188.

7

Estimating the Healthy Life Expectancy (HLE) in the Far Past: The Case of Sweden (1751–2016) with Forecasts to 2060

Healthy life expectancy (HLE) estimates are achieved after systematic work done by a large group of researchers all over the world over the last few decades. The most successful estimate was termed as HALE and is provided by the World Health Organization (WHO) on their related website. Having established a methodology of data collection and handling, the HLE can be estimated and provided for researchers and policy makers.

However, there remains an unexplored period over the last few centuries where LE (life expectancy) data exists along with the appropriate life tables, but not enough information for HLE estimates is collected and stored. The problem has now been solved following a methodology of estimating the HLE from the life tables after the healthy life years lost (HLYL) estimation.

Our methodology on a Direct HLYL estimation from life tables is tested and verified via a series of additional methods including a Weibull parameter test, a Gompertz parameter alternative and of course a comparison with HALE estimates from the WHO.

Chapter written by Christos H. SKIADAS and Charilaos SKIADAS.

7.1. Life expectancy and healthy life expectancy estimates

The full life tables are used to estimate not only the LE but also the HLE, based on an existing methodology (Skiadas and Skiadas 2020a).

Based on the data series from 1900 to 2016 for males and females in Sweden, estimates until 2016 and forecasts to 2060 are done. The logistic model is fitted to data series to calculate the three parameters of the model. Then, forecasts to 2060 are done. For fitting and long range forecasts, the logistic model is selected.

1900 was a milestone in health improvement in many countries. As Jan Sundin and Sam Willner (2007) report for Sweden:

> Some diagnoses, such as smallpox, ague, dysentery and cholera, had essentially disappeared in 1900, whilst diphtheria, whooping cough and measles continued to be a real, albeit drastically reduced, threat during childhood.

According to Sunding and Willner, in the years just before 1900, several important key-points have been done in Sweden including:

– 1862: local government reform: establishment of the Landsting (county councils, who take over the responsibility for hospitals);

– 1878: the National Medical Board (Medicinalstyrelsen) is founded;

– 1890: a chief provincial doctor is appointed in every county;

– 1891: the first sanatorium for lung disease sufferers in Sweden is opened.

The development of the first healthcare system of modern history started with policies introduced by Otto von Bismarck's social legislation (1883–1911). The introduction of such systems in many countries came after important discoveries from scientists such as Pasteur, Chamberland and Descomby in France, von Behring in Germany, Kitasato from Japan and many others. The 1901 Nobel Prize in Physiology or Medicine, the first one in that field, was awarded to von Behring for his discovery of a diphtheria antitoxin.

It looks like the healthcare systems and methodologies already set in place in 1900 follow a rather systematic trend until today. See Figure 7.1 where LE data series is provided by the human mortality database (HMD), and HLE is estimated with our direct methodology (Skiadas and Skiadas 2018a, 2018b, 2020a, 2020b, 2020c). The LE series from 1751 to 1875 is strongly fluctuating mainly due to health causes. The fluctuations become smaller after this period, with a clear stabilization from 1900 until now except the strong decline during the 1918 influenza pandemic followed by a fast recovery later on. The period starting from 1950 is followed with a rather smooth trend as a result of the improvement of the health system structure, financing, technology and pharmaceutical discoveries and production.

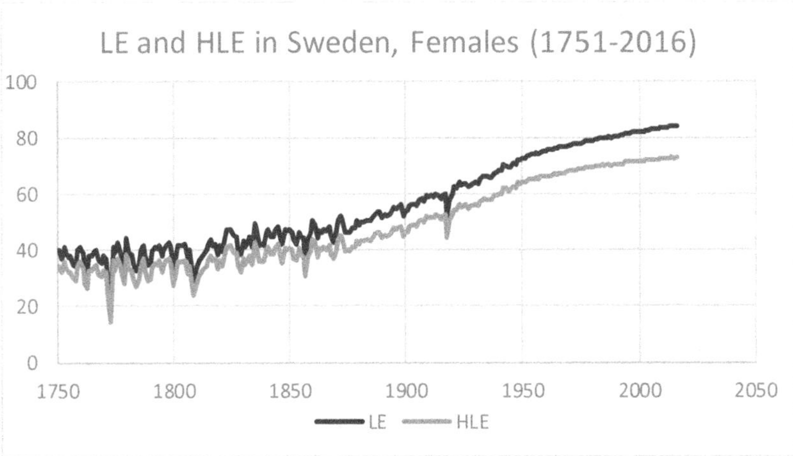

Figure 7.1. *Life expectancy (LE) and healthy life expectancy (HLE) in Sweden, females (1751–2016). For a color version of this figure, see www.iste.co.uk/dimotikalis/analysis2.zip*

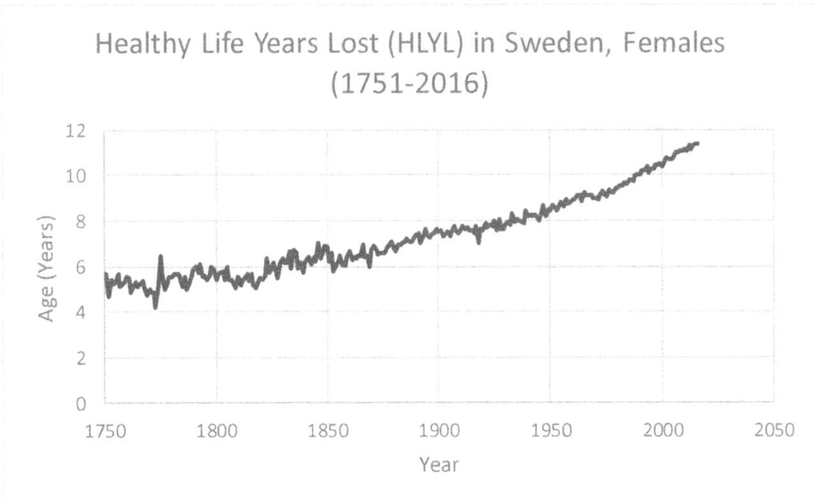

Figure 7.2. *Healthy life years lost (HLYL) in Sweden, females (1751–2016)*

The healthy life years lost (HLYL) calculated data series is illustrated in Figure 7.2. The HLYL trend grows slightly until 1850 followed by a faster growth until 11.35 years of age in 2016, after 5.69 years of age in 1751.

7.2. The logistic model

This classical model proposed by P.F. Verhulst in 1838 to estimate the population of France is proven to be a successful tool for long range forecasting. In his first application, Verhulst predicted the population of France for almost 100 years. Pearl and Reed used this model to predict the growth of the United States' population. Applications in other countries have also been done.

The three parameter logistic model equation form is the following:

$$g(t) = \frac{F}{1 + \left(\frac{F}{g(0)} - 1\right) exp(-b(t - T(0)))}$$

where b is the trend or diffusion parameter, F is the upper level of the sigmoid logistic process and g(0) is the value at time T(0) = 1900.

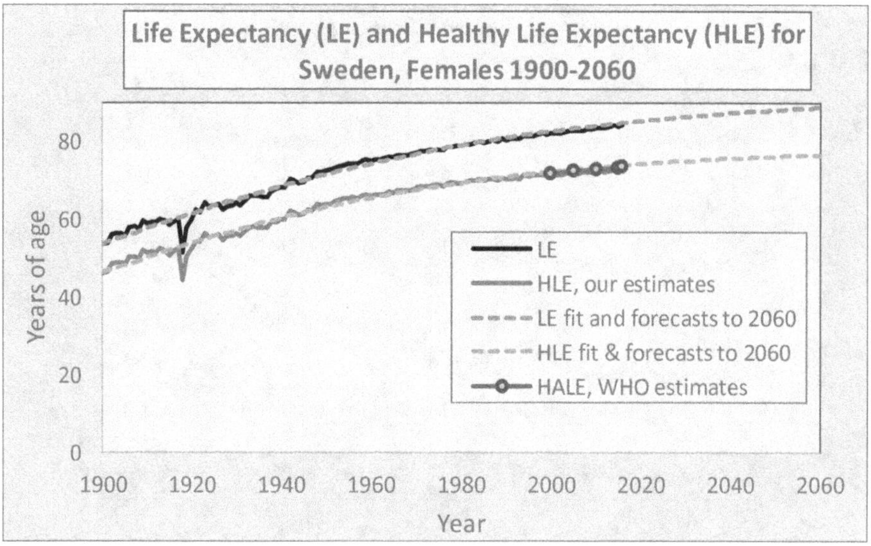

Figure 7.3. *Logistic model fit and forecasts to 2060 for females in Sweden. For a color version of this figure, see www.iste.co.uk/dimotikalis/analysis2.zip*

7.3. The HALE estimates and our direct calculations

The latest WHO estimates for healthy life expectancy called HALE are provided for the years 2000, 2005, 2010, 2015 and 2016.

These estimates perfectly fit into our calculations for the HLE and the fit results by using the logistic model.

Our HLE calculations are based on the direct estimates from the life tables of the HLYL with a formula provided in recent publications (Skiadas and Skiadas 2020a, 2020b, 2020c) that is:

$$HLYL = max\frac{xm_x}{\sum_0^x m_x}$$

where m_x is the mortality at age x as provided in HMD life tables.

Then, HLE = LE-HLYL.

The logistic model is applied to data sets for LE and HLE from our estimates from 1900 to 2016. The parameters selected appear in Table 7.1.

The HLYL are 7.40 years of age in 1900, 10.19 in 2016 and 12.15 in 2060 with a maximum of F = 13.74 years of age difference.

	Logistic model parameters		LE and HLE in 1900, 2016 and 2060		
	b	F	1900	2016	2060
LE	0.01820	92.15	53.70	84.09	88.70
HLE	0.02096	78.41	46.30	73.90	76.55
HLYL = LE-HLE		13.74	7.40	10.19	12.15

Table 7.1. *Logistic model parameters and estimates*

Table 7.2 summarizes the three HLE estimates from the WHO (HALE), our direct estimates and from logistic fit. All three methodologies provide close results.

Year	2000	2005	2010	2015	2016
WHO HALE	71.74	72.37	72.97	73.27	73.36
Direct HLE	71.63	71.98	72.32	72.67	72.74
Logistic fit	72.25	72.81	73.34	73.81	73.90

Table 7.2. *HALE and healthy life expectancy direct estimates and logistic fit*

7.4. Conclusion

We have solved the problem of finding the HLE in the far past. The case of Sweden (1751–2016, females) with forecasts to 2060 and comparisons with HALE have been explored. The selected logistic model has a good fit, while the HALE estimates from the WHO compared very well to our estimates both with direct method and the logistic fit.

7.5. References

Skiadas, C.H. and Skiadas, C. (2018a). *Exploring the Health State of a Population by Dynamic Modeling Methods*. Springer, Cham, Switzerland.

Skiadas, C.H. and Skiadas, C. (2018b). *Demography and Health Issues: Population Aging, Mortality and Data Analysis*. Springer, Cham, Switzerland.

Skiadas, C.H. and Skiadas, C. (2020a). *Demography of Population Health, Aging and Health Expenditures*. Springer, Cham, Switzerland [Online]. Available at: https://www.springer.com/gp/book/9783030446949.

Skiadas, C.H. and Skiadas, C. (2020b). *Relation of the Weibull Shape Parameter with the Healthy Life Years Lost Estimates: Analytical Derivation and Estimation from an Extended Life Table*. Springer, Cham, Switzerland.

Skiadas, C.H. and Skiadas, C. (2020c). *Direct Healthy Life Expectancy Estimates from Life Tables with a Sullivan Extension. Bridging the Gap Between HALE and Eurostat Estimates*. Springer, Cham, Switzerland.

Sundin, J. and Willner, S. (2007). *Social Change and Health in Sweden: 250 Years of Politics and Practice*. Swedish National Institute of Public Health, Solna, Sweden.

8

Vaccination Coverage Against Seasonal Influenza of Workers in the Primary Health Care Units in the Prefecture of Chania

Influenza is a contagious disease of the respiratory system, which causes mild to severe disease and leads to disorganization in both professional and social lives. The most effective way to minimize its spread is through the influenza vaccine, which protects the population from its crucial complications. The purpose of this dissertation is to record the vaccination coverage rate in the last three years and search for factors that can affect the vaccination positively or negatively. The research is based on data obtained from the questionnaires distributed to the Health Centers (HC) and Local Health Units (LHU) of the prefecture of Chania, in the period February to March 2020, and compared with data from studies of both Greek and international scientific literature over the last 15 years. The percentage of influenza vaccinations of employees in public structures PHC in the prefecture of Chania shows an increasing trend in the last three years, with 41.7% in 2017–2018, 48.4% in 2018–2019 and 57.1% in 2019–2020. Men are vaccinated more often than women, with the highest rates recorded at the age of ≤ 34 years, while in the period 2019–2020, 68.7% of doctors, 57.6% of nurses and 43.6% other staff were vaccinated. The most common reasons for avoiding vaccination are the view that there is no risk of influenza (29.4%), the fear of vaccine safety/side effects (25.5%) and the non-vaccination due to inactivity (19.6%). The three main reasons for motivating staff to have the vaccination are the need for family protection (65.3%), the need for self-protection (63.7%) and immunization due to work. The role of the administration and the scientific community is vital in order to improve the rates of influenza vaccination and alter the mentality in favor of prevention. Equally important, however, is the role of workers, as they need to consider the potential

Chapter written by Aggeliki MARAGKAKI and George MATALLIOTAKIS.

consequences of transmitting the influenza to health care facilities and support vaccination as a key means of protecting themselves and patients.

8.1. Introduction

Influenza is a contagious disease of the respiratory system, which causes mild to severe illness and can even lead to death. The most effective way to prevent it is the annual influenza vaccine, which protects against the transmission of the virus and its serious complications (Ministry of Health 2019a, 2019b, 2019c).

The rationale for the influenza vaccination is based on the need to protect the health of workers and vulnerable patients, as well as ensuring the proper functioning of health services during influenza seasons. Acceptance by health care staff helps build confidence in vaccination and better prepare the health system for the next influenza pandemic (Ministry of Health 2019a, 2019b, 2019c).

Unvaccinated health professionals are the main source of influenza transmission in several hospital outbreaks, so staff immunization acts as a barrier against its spread. The percentage of the annual influenza vaccinations of health units is an indicator of the compliance of controlling the spread of the virus. It evaluates the efficiency of the administration and the quality of the provided health services. The Committee for Nosocomial Infections plays an important role, promoting and facilitating the vaccination of staff while also gathering data on vaccination coverage in order to adequately monitor the influenza (Ministry of Health 2019a, 2019b, 2019c; National Public Health Organization 2020).

8.2. Material and method

The sample of respondents comes from medical and other staff in Health Centers (HC) and Local Health Units (LHU) in the prefecture of Chania, from February to March 2020. 80% of the questionnaires were answered. Of these, 63 people (40.4%) work in the first and second Chania LHU and the first and second Chania HC (structures within the city of Chania), while the remaining 93 (59.6%) work in the Kissamos HC, the Kandanos HC and the Vamos HC (structures outside the city of Chania) (Figure 8.1).

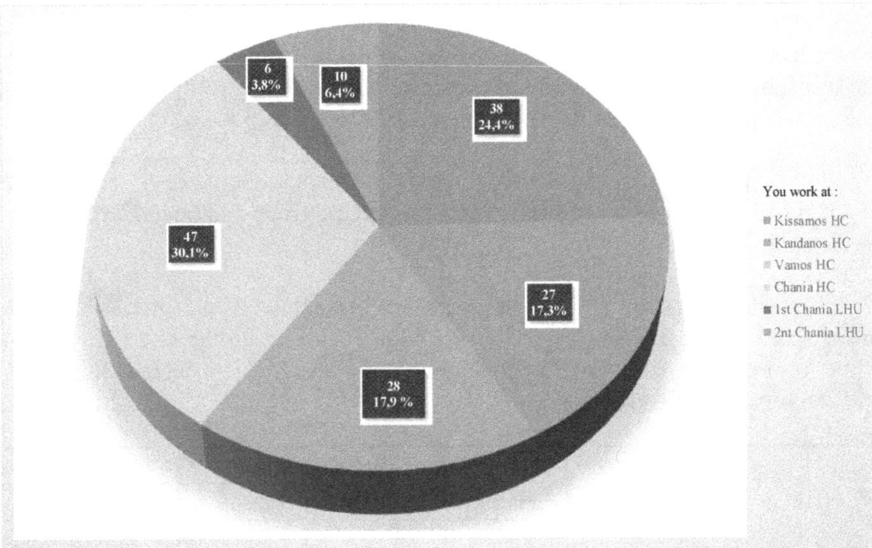

Figure 8.1. *Participation of respondents from the Health Centers and Local Health Units of the prefecture of Chania*

In all participants, women made up 69.2% of the sample, with the largest age group being 45–54 years (52, 33.3%), and the smallest being 55+ (27, 17.3%). Regarding the distribution, 59 women constituted 63.4% of the sample in the out of town HC/LHU, while 49 women constituted 77.8% in the inner city, with the gender distribution not showing a statistically significant difference (p = 0.057) (Table 8.1).

		HC/LHU location						
		Inside city (93)		**Outside city (63)**		**Total**		
		n	**%**	**n**	**%**	**n**	**%**	**p**
Gender	Male	34	36.60	14	22.20	48	30.80	0.057
	Female	59	63.40	49	77.80	108	69.20	
Age group	<=34	24	25.80	14	22.20	38	24.40	0.351
	35–44	20	21.50	19	30.20	39	25.00	
	45–54	35	37.60	17	27.00	52	33.30	
	55+	14	15.10	13	20.60	27	17.30	

Table 8.1. *Age and gender distribution in terms of HC/LHU inside/outside the city*

A total of 67 respondents (43.2%) in the sample were medical staff, while 33 respondents were nurses (21.3%). There was a statistically significant difference in the distribution of doctors, with those outside the city being 49 (52.7%) and those within the city being 18 (29.0%) of the respective samples ($p = 0.011$).

Work experience showed that in total (inside/outside the city), 85 respondents (54.5%) had worked up to 15 years, while the largest group had one to five years of service (29, 18.6%).

In terms of education, the largest percentage had a university degree (84, 53.8%) in total and also at HC/LHU outside (51, 54.8%) and inside the city (33, 52.4%), while from out of the total of 25 people who had a postgraduate degree, 20 (80.0%) had a master's degree (Table 8.2).

		HC/LHU location						
		Inside city (93)		**Outside city (63)**		**Total**		
		n	%	n	%	n	%	p
Staff	Medical	49	52.70	18	29.00	67	43.20	0.011
	Nurses	18	19.40	15	24.20	33	21.30	
	Others	26	28.00	29	46.80	55	35.50	
Work experience (years)	<1	11	11.80	5	7.90	16	10.30	0.08
	1–5	15	16.10	14	22.20	29	18.60	
	6–10	16	17.20	5	7.90	21	13.50	
	11–15	12	12.90	7	11.10	19	12.20	
	16–20	11	11.80	6	9.50	17	10.90	
	21–25	11	11.80	3	4.80	14	9.00	
	26–30	4	4.30	10	15.90	14	9.00	
	>30	13	14.00	13	20.60	26	16.70	
Education	High school	14	15.10	4	6.30	18	11.50	0.225
	Occupational training institute	10	10.80	7	11.10	17	10.90	
	Technological Educational institute	18	19.40	19	30.20	37	23.70	
	University	51	54.80	33	52.40	84	53.80	
Postgraduate degree	Master	4	57.10	16	88.90	20	80.00	0.075
	PhD	3	42.90	2	11.10	5	20.00	

Table 8.2. *Professional characteristics regarding HC/LHU inside/outside the city*

8.3. Results

The vaccination coverage in total for the periods 2017–2018, 2018–2019 and 2019–2020 is presented in Figure 8.2. There is an increasing trend in the vaccination of employees with 65 respondents (41.7%) for 2017–2018, 74 respondents (48, 4%) for 2018–2019 and 89 respondents (57.1%) for 2019–2020. Also, three respondents (1.9%) chose I do not know/I do not want to answer for the vaccination 2017–2018, four (2.6%) for the vaccination 2018–2019 and none for 2019–2020.

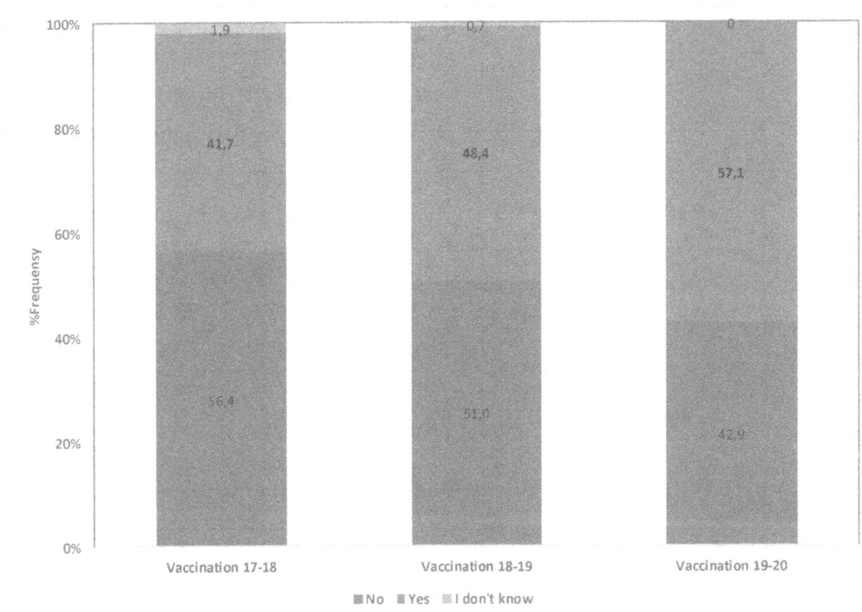

Figure 8.2. *Self-reported frequency of influenza vaccinations 2017–2020. For a color version of this figure, see www.iste.co.uk/dimotikalis/analysis2.zip*

The frequency of vaccinations and 95% CI in terms of the total number of respondents, their age distribution and gender is presented in Table 8.3. Based on the results in all three time periods 2017–2018, 2018–2019 and 2019–2020, men have a higher frequency of vaccination, with respective values of 50.0% (36.2%–63.8%), 59.6% (45.3%–72.7%) and 64.6% (50.5%–76.9%). Women for the respective time periods had % frequency with 95% CI, 38.0% (29.2%–47.3%) for 2017–2018, 43.4% (34.2%–52.9%) for 2018–2019 and 53.7% (44.3%–62.9%) for 2019–2020. Also comparing the vaccination rates by gender, no statistically significant difference was observed in 2017–2018 ($p = 0.159$), 2018–2019 ($p = 0.065$) or 2019–2020 ($p = 0.205$).

In all three study periods, the highest vaccination rates were observed at the age <= 34 years (2017–2018: 50.0%, 2018–2019: 63.9%, 2019–2020: 76.3%); however, there was no statistically significant difference in the periods 2017–2018 ($p = 0.399$) and 2018–2019 ($p = 0.132$), but only in the period 2019–2020 ($p = 0.012$).

	Vaccine 2017–2018	**n**	**% n**	**95% LL**	**95% UL**	**p**
Age	<=34	19	50.00	34.60	65.40	0.399
	35–44	17	43.60	28.90	59.10	
	45–54	17	32.70	21.10	46.10	
	55+	12	44.40	27.10	62.90	
Gender	Male	24	50.00	36.20	63.80	0.159
	Female	41	38.00	29.20	47.30	
Total		65	41.70	34.10	49.50	
	Vaccine 2018–2019	**n**	**% n**	**95% LL**	**95% UL**	**p**
Age	<=34	23	63.90	47.60	78.00	0.132
	35–44	18	46.20	31.30	61.60	
	45–54	20	38.50	26.20	52.00	
	55+	13	50.00	31.60	68.40	
Gender	Male	28	59.60	45.30	72.70	0.065
	Female	46	43.40	34.20	52.90	
Total		74	48.40	40.50	56.30	
	Vaccine 2019–2020	**n**	**% n**	**95% LL**	**95% UL**	**p**
Age	<=34	29	76.30	61.20	87.60	0.012
	35–44	21	53.80	38.40	68.70	
	45–54	22	42.30	29.60	55.80	
	55+	17	63.00	44.20	79.10	
Gender	Male	31	64.60	50.50	76.90	0.205
	Female	58	53.70	44.30	62.90	
Total		89	57.10	49.20	64.60	

Table 8.3. *% frequency and 95% CI of vaccinations in total and by gender, age of vaccination periods from 2017 to 2020*

Vaccinations were also studied, based on whether they belonged to HC/LHU inside/outside the city and in terms of the status of the respondents. Table 8.4 presents the results of HC/LHU inside/outside the city. Vaccination rates do not differ inside/outside the city for 2017–2018 (50.8%, 38.6%–62.9%)/(35.5%, 26.3%–45.5%) with p = 0.057, as well as for 2019–2020 (65.1%, 52.8%–76.0%)/(51.6%, 41.5%–61.6%) with p = 0.095. For the period 2018–2019, the vaccinations were statistically higher with a percentage of 61.3% (48.9%–72.7%) in the city, with p = 0.008 compared to those outside the city 39.6% (30.0%–49.8%).

	Vaccine 17–18	**n**	**% n**	**95% LL**	**95% UL**	**p**
City	Inside	33	35.50	26.30	45.50	0.057
	Outside	32	50.80	38.60	62.90	
	Vaccine 18–19	**n**	**% n**	**95% LL**	**95% UL**	**p**
City	Inside	36	39.60	30.00	49.80	0.008
	Outside	38	61.30	48.90	72.70	
	Vaccine 19–20	**n**	**% n**	**95% LL**	**95% UL**	**p**
City	Inside	48	51.60	41.50	61.60	0.095
	Outside	41	65.10	52.80	76.00	

Table 8.4. *Frequency of vaccinations and 95% CI between HC/LHU inside/outside the city*

Figure 8.3 shows the % vaccination frequency for the period 2017–2020 in all three occupational categories of participants. Doctors show a similar frequency of 52.2% (40.4%–63.9%) with nurses 51.5% (34.9%–67.8%) for the period 2017–2018, while the vaccination frequency for the rest of the staff was low, at 23.6% (13.9%–36.0%). There was a statistically significant difference between the above percentages (p = 0.003).

For the period 2018–2019 despite the fact that the frequency of vaccination of nurses increased to 60.6% (43.6%–75.8%) and the rest of the staff to 35.8% (24.0%–49.2%), the frequency of vaccination of doctors remained relatively constant with the previous period 53.0% (41.4%–64.7%). No statistically significant difference was observed (p = 0.053).

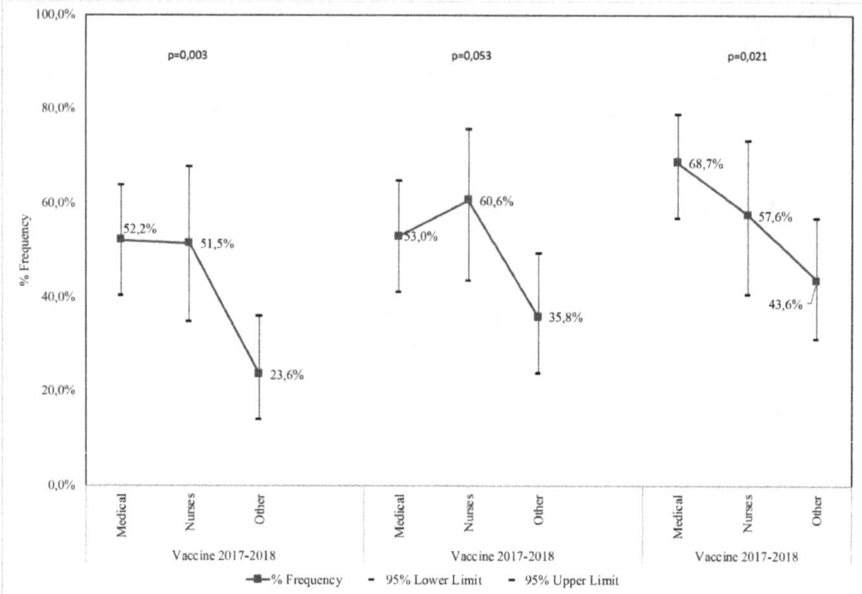

Figure 8.3. *% frequency and 95% CI of influenza vaccination among staff*

For the period 2019–2020, there was a statistically significant difference ($p = 0.021$) between the vaccination frequencies of doctors 68.7% (56.9%–78.8%), nurses 57.6% (40.7%–73.2%) and the remaining staff 43.6% (31.1%–56.8%).

Table 8.5 presents the distribution by the type of staff and in total, of the "impulses" (motivations, measures, views) and also the reasons for prevention of vaccination. Family protection (96, 65.3%) and the need for self-protection (93, 63.7%) are the two most common "impulses" recorded for vaccination, with no significant variations between the types of staff. A statistically significant difference ($p = 0.037$) occurred in the "need to protect patients", where the rest of the staff responded positively at a rate of 30.0% ($n = 15$).

Finally, in terms of vaccine prevention, the most common are fears about the safety of the vaccine and its side effects (53, 38.7%), inactivity for vaccination (35, 25.5%) and the belief that the respondent will not become ill (29, 21.2%). Three of the remaining staff (6.5%) were "anti-vaccinators" which varied the results ($p = 0.048$). Another statistical differentiation with $p = 0.005$ was observed regarding the availability of the vaccine in the workplace in HC/LHU (eight doctors 13.1%).

Impulses – motivations	Staff								
	Medical		Nurses		Other		Total		
	n	%	n	%	n	%	n	%	p
Encouragement from work	18	27.30	8	26.70	7	14.00	33	22.60	0.2
Need for patient protection	35	53.00	15	50.00	15	30.00	65	44.50	0.037
Need for self-protection	47	71.20	18	60.00	28	56.00	93	63.70	0.215
Work-related immunization	35	53.00	16	53.30	22	44.00	73	50.00	0.578
Family Protection	44	65.70	19	63.30	33	66.00	96	65.30	0.967
Free Vaccination	11	16.70	6	20.00	5	10.00	22	15.10	0.426
Preventions	Staff								
	Medical		Nurses		Other		Total		
	n	%	n	%	n	%	n	%	p
Belief that I will not get sick	13	21.30	8	26.70	8	17.40	29	21.20	0.626
I do not consider it effective	6	9.80	2	6.70	8	17.40	16	11.70	0.303
I do not consider the influenza serious	3	5.00	4	13.30	4	8.70	11	8.10	0.386
It does not help in prevention	5	8.20	4	13.30	7	15.20	16	11.70	0.508
Lack of sufficient information	9	14.80	8	26.70	6	13.00	23	16.80	0.254
Fear of safety/side effects	21	34.40	10	33.30	22	47.80	53	38.70	0.294
Inaction	21	34.40	5	16.70	9	19.60	35	25.50	0.098
I am against vaccinations	0	0.00	0	0.00	3	6.50	3	2.20	0.048
Not available in my work place	8	13.10	0	0.00	0	0.00	8	5.90	0.005

Table 8.5. *Breakdown by type of staff of impulses and preventions of vaccination*

8.4. Discussion

Similar surveys have been conducted in primary and secondary health facilities in Greece and abroad.

In a study by Maltezou *et al.* (2007) for the period 2005–2006, conducted in 132 hospitals in Greece, the influenza vaccination rates of health workers were 16.36%, when in the previous year, they were only 1.72%. In 2006–2007 (Maltezou *et al.* 2008), the average influenza vaccination was 5.8%, in which 89.1% did so in order to protect themselves, 59.1% to protect their family and 55.2% their patients. The main reasons for refusing vaccination were the perceptions about non-disease (43.2%), about the ineffectiveness of the vaccine (19.2%) and the fear of its side effects (33.4%). In 2009, researchers (Maltezou *et al.* 2010) recorded the intention of health professionals in 92 hospitals and 60 Health Centers in Greece for influenza vaccination. 21.8% stated that they intended to be vaccinated, while the main reasons for refusing vaccination were the fear of the safety of the vaccine (43.1%), insufficient information (27.8%) and the perception that there is no risk of influenza (10.7%).

Regarding studies abroad (Dominguez *et al.* 2013), the rate of influenza vaccination of workers in Spanish PHC units in 2011–2012 was 50.7%. Factors that positively influenced the decision to vaccinate were the risk of occupational disease, protection against influenza and its complications, and the view that the vaccine is important for health care workers.

In the study conducted by Durando *et al.* (2016) in Italy, during the period 2013–2014, almost half of the study population had never been vaccinated against influenza between 2008 and 2014. In the study conducted by Petek and Kamnik-Jug (2018) in PHC units in Slovenia, only 12% of health professionals were vaccinated during the period 2014/2015. Motivation for vaccination was the fear of risk of infection in the workplace, self-protection and the protection of family and colleagues. The main obstacles were the doubt about the effectiveness of the vaccine, the fear of side effects and the belief that they are not at high risk of infection from influenza.

In the study conducted by Yu *et al.* (2019), only 6% (out of 4,706) of nurses in China got the seasonal influenza vaccine during 2017/2018 due to lack of time (28%) and confidence in its effectiveness (12%).

Finally, according to National Public Health Organization (NPHO) and the aggregate results of the period 2019–2020, the vaccination coverage of the staff of health services nationwide in the PHC (HC and LHU) was 57.9%. In the period 2018–2019, the corresponding percentages were 43.7%; in 2017–2018, it was 40.2%; in 2016–2017, it was 34.6%; and in 2015–2016, it was 24.3%. Regarding the anti-influenza coverage of the staff of the public structures of PHC of the

Administration of the Seventh Health District (AHD of Crete), to which the corresponding structures of the prefecture of Chania belong, in the period 2019–2020, the percentage of influenza vaccination was 65% and in 2018–2019, it was 59.4% (National Public Health Organization 2020).

The research showed that the percentage of vaccination coverage against the seasonal influenza of PHC employees in the prefecture of Chania shows an increasing trend in the last three years. However, there is room for improvement, so health structures should implement policies to encourage and promote influenza vaccination. The role of the administration and scientific community is important in order, through proper guidance and education, to change the mentality in favor of prevention and overthrow the dangerous anti-vaccination culture that was based on the phobia about the possible side effects of the vaccine.

8.5. References

Dominguez, A., Godoy, P., Castilla, G. (2013). Knowledge of and attitudes to influenza vaccination in healthy primary healthcare workers in Spain, 2011–2012. *PLoS ONE*, 8(11), e81200.

Durando, P., Alicino, C., Dini, G. (2016). Determinants of adherence to seasonal influenza vaccination among healthcare workers from an Italian region: Results from a cross-sectional study. *BMJ Open*, 6(5), e010779.

Jianxing, Y., Xiang, R., Chuchu, Y. (2019). Influenza vaccination coverage among registered nurses in China during 2017–2018. An Internet Panel Survey. *Vaccines*, 7(4), 134.

Maltezou, H.C., Maragos, A., Halharapi, T. (2007). Factors influencing influenza vaccination rates among healthcare workers in Greek hospitals. *Journal of Hospital Infection*, 66(2), 156–159.

Maltezou, H.C., Maragos, A., Katerelos, P. (2008). Influenza vaccination acceptance among health-care workers: A nationwide survey. *Vaccine*, 26(11), 1408–1410.

Maltezou, H.C., Dedoukou, X., Patrinos, S. (2010). Determinants of intention to get vaccinated against novel (pandemic) influenza A H1N1 among health-care workers in a nationwide survey. *Journal of Infection*, 61(3), 252–258.

Ministry of Health (2019a). *Influenza vaccination of health care personnel* [Online]. Available at: https://eody.gov.gr/wp-content/uploads/2019/01/antigripikos-emvoliasmos-prosopikou-yy.pdf [Accessed April 2020].

Ministry of Health (2019b). Instructions for seasonal influenza 2019–2020. Influenza vaccination [Online]. Available at: https://www.moh.gov.gr/articles/health/dieythynsh-dhmosias-ygieinhs/metadotika-kai-mh-metadotika-noshmata/c388-egkyklioi/6474-odhgies-gia-thn-epoxikh-griph-2019-2020-ndash-antigripikos-emboliasmos?fdl=15434 [Accessed April 2020].

Ministry of Health (2019c). Influenza vaccination action plan [Online]. Available at: https://eody.gov.gr/wp-content/uploads/2019/01/antigripikos-emvoliasmos-prosopikou-yy.pdf [Accessed April 2020].

National Public Health Organization (2020). *Influenza vaccination of health services personnel influenza season 2019–2020* [Online]. Available at: https://eody.gov.gr/wp-content/uploads/2020/05/ekthesi_emvoliasmos_ergazomenon_gripi_2019-2020.pdf [Accessed June 2020].

Petek, D. and Kamnik-Jug, K. (2018). Motivators and barriers to vaccination of health professionals against seasonal influenza in primary healthcare. *BMC Health Services Research*, 18(1), 853.

9

Some Remarks on the Coronavirus Pandemic in Europe

Severe acute respiratory syndrome coronavirus 2 (SARS-CoV-2) was initially reported in China in late 2019 and rapidly spread across the world. On March 11, 2020, the World Health Organization (WHO) characterized the situation as a global pandemic. By late April 2020, there were 2,475,440 confirmed cases and 170,069 deaths worldwide; 164,656 of the fatal cases were reported in Europe. In response to the pandemic, the primary objectives of European countries were: (1) to limit the spread of the virus, (2) to protect the most vulnerable and health workers and (3) to provide a clear quantification of the virus's process. However, there are major inconsistencies between countries concerning the adoption of protocols and measures towards the above goals. The installation of measurements, such as restrictions on public gatherings, lockdowns, and the shutdown of educational institutions and workplaces, took place at different times and on different scales. In parallel, there are various issues and restrictions concerning the epidemiological data we possess thus far, a fact that hinders the study of this pandemic.

9.1. Introduction

Throughout its existence, humanity has encountered various plagues, epidemics and pandemics. It was by the late 20th century, and the adoption of modern virology and related disciplines, that we became able to track down and to analyze the menacing pathogens thoroughly. The three major worldwide influenza outbreaks of our previous century were ascribed to H2N2, H3N2 and to H1N1 viruses (Kilbourne 2006; Liu *et al.* 2018; Schwartz 2018; Honigsbaum 2020). The 21st century has also

---

Chapter written by Konstantinos ZAFEIRIS and Marianna KOUKLI.

been characterized by the emergence of fatal viruses that led to both epidemics and pandemics such as MERS-CoV (Middle East respiratory syndrome coronavirus) and SARS-CoV (severe acute respiratory syndrome coronavirus) and the Ebola virus. Their most prominent common characteristics were the large-scale outbreaks that followed their initial emergence, as well as both the high fatality and basic reproduction rates (R_0) (Callaway *et al.* 2020; Kaswa and Govender 2020; Petersen *et al.* 2020).

In the early months of 2020, humanity was rocked by the emergence of the novel severe acute respiratory syndrome coronavirus 2 (SARS-CoV-2) that initially induced various types of pneumonia but that ultimately led to a severe public health crisis worldwide. This is despite the fact that, according to the epidemiological data gathered thus far, the latter virus, responsible for COVID-19, presents lower fatality and R_0 rates (i.e. the estimated number of new infections from a single case) compared to the former pathogens. Its severity therefore lies in the fact that it seems to be able to spread very easily (Abebe *et al.* 2020; Callaway *et al.* 2020; Khalili *et al.* 2020). Soon after its detection, COVID-19 was characterized as a highly contagious disease, leading to a global effort to investigate it in order to inhibit its spread.

9.2. Background

9.2.1. *CoV pathogens*

The COVID-19 virus belongs to a family of coronaviruses (CoV), first characterized and described in the mid-20th century. They form a large family of enveloped, single-stranded, positive-strand RNA viruses, and owe their name to their corona-shaped morphology. Studies on the biochemical and molecular nature of coronaviruses have shown that they comprise of several pathogens that can infect both humans and animals, causing acute and chronic respiratory, gastroenteric diseases, as well as problems in the central nervous system of the host and hepatitis (Lai and Holmes 2001; Dasari and Bhukya 2020; Malik *et al.* 2020).

Coronaviruses present a distinctive genetic characteristic among all other viruses, which is also related to their variability: they contain the largest genomes and thus an expansive coding capacity, which equates to high variability in gene expression (Weiss 2005; Masters 2006; Tort *et al.* 2020). In 2002–2003, the interest of the scientific community was drawn to the emergence of a new viron pathogen, a coronavirus that provoked an intense outbreak of a severe acute respiratory syndrome (SARS-CoV). The particular characteristics of the SARS-CoV genome sequence have raised many questions as to its taxonomic classification. Therefore, research on the phylogenetic analysis of viral proteins indicated that SARS-CoV is a distant member of the second group of coronaviruses (Weiss and Martin 2005;

Masters 2006; Malik *et al.* 2020), whereas some scholars acknowledge it as a member of a new fourth group of coronaviruses (Marra 2003; Rota 2003). The 2002–2003 epidemic originated in the Guangdong Province in China and led to a total number of 8,000 cases and 800 deaths (approximately a 9.5% mortality rate) worldwide (Schoeman and Fielding 2019; Agarwal and Agarwal 2020; Prajapati 2020; Tort *et al.* 2020). Furthermore, in 2012 highly elevated mortality rates (approximately 35%) were reported in the successive MERS-CoV epidemic (Zaki *et al.* 2012; Jang and Seo 2020).

In late 2019, a third epidemic caused by a novel coronavirus – SARS-CoV-2 – turned into a pandemic. It first emerged in Wuhan, in the Hubei Province of China, causing what was described as atypical pneumonia, followed by a severe acute respiratory infection. It rapidly became clear that this novel pathogen, initially found in bats, was a major threat to humans all over the world.

9.2.2. *Clinical characteristics of COVID-19*

It has been suggested that the angiotensin-converting enzyme II (ACE-2) works as a receptor of the virus (Chan *et al.* 2020; Zhou *et al.* 2020). Entry into the host cell is achieved by a glycoprotein, called "spike" (S), found within the protein shell of the virus that connects to the ACE-2 receptor (Hoffmann *et al.* 2020; Malik *et al.* 2020; Prajapati *et al.* 2020). Immediately after binding to the host cell, the virus provokes a rapid infection that leads to drastic changes. The envelope (E) protein of the CoV pathogen is related to viral morphogenesis and assembly, whereas the membrane (M) protein enables RNA transcription of the virus. However, one of the main features of CoV is that it can protect itself using the host immune system due to its ability to adapt to the host environment and to mutate (Neuman *et al.* 2012; Ruch *et al.* 2012; Sui *et al.* 2014; Li 2016; Dasari and Bhukya 2020; Lu *et al.* 2020).

Lungs contain a large number of the ACE-2 enzyme and, therefore, can be very vulnerable to a SARS-CoV-2 infection (Prajapati *et al.* 2020). This might be the reason why, in the case of respiratory system, infections are the most common. For example, according to worldwide studies, the largest percentage of patients present the following: dry cough, shortness of breath, intense dyspnea and tachypnea, sputum, as well as fever (Guan *et al.* 2020; Huang *et al.* 2020; Tabata *et al.* 2020; Wang *et al.* 2020; Zhang *et al.* 2020). In addition, headache, loss of smell, nasal obstruction, rhinorrhea and sore throat have also been reported (European Center for Disease Prevention and Control 2020).

In addition to pulmonary symptoms, a significant number of COVID-19 patients also suffer from gastrointestinal problems. The responsible biological mechanisms remain unknown; however, it must be mentioned that the frequency of such symptoms is very high. More specifically, various research works have shown that the presence of gastrointestinal symptoms occurs within 20% to 50% of COVID-19 cases (Pan *et al.* 2020; Ye *et al.* 2020; Wang *et al.* 2020). Symptoms like diarrhea and loss of appetite are among the most common, followed by vomiting and abdominal pain (Manabe *et al.* 2020; Rajendran *et al.* 2020).

Furthermore, cardiovascular problems in COVID-19 patients have also manifested. Alterations in the cardiac troponin-I level have been reported as being linked to myocardial injuries. Arrhythmia, chest pain, hypoxemia and lymphopenia are some of the most commonly reported symptoms (Gulati *et al.* 2020; Momtazmanesh *et al.* 2020; Zheng *et al.* 2020).

Neurological problems have also been reported in various cases. Issues related to the sensory system, such as anosmia, hypogeusia and dysgeusia are among the most common (Aghagoli *et al.* 2020; Fiani *et al.* 2020). Fewer, but rather more severe, are the cases that involve paresthesia, altered mental status and encephalopathy which so far appear to be associated with previous health problems (Demirci Otluoglu *et al.* 2020; Poyiadji *et al.* 2020).

From a broader perspective, according to various case studies, there is a plethora of symptoms that cannot be fully understood or linked to a clear biological mechanism. Myalgia, liver dysfunction and hematological problems, such as thrombosis and hemostasis, have also been reported, even though they appear in a low number of cases (Hardenberg and Luft 2020; Liao *et al.* 2020; Nepal *et al.* 2020). It must be mentioned, however, that the degree of severity as well as the evolution of the disease varies, and it is linked to various factors such as the medical history of the patient and the level of medical care they receive. In general, in the case of COVID-19, there is a high risk of aggravation, when the atypical pneumonia may lead to acute respiratory distress syndrome (ARDS), the gastrointestinal and cardiovascular problems may result in severe damage to the intestinal mucosal barrier and to acute cardiac injuries (Guan *et al.* 2019; Abdin *et al.* 2020; Fang *et al.* 2020; Gulati *et al.* 2020; Gupta *et al.* 2020; Ktreitmann *et al.* 2020; Momtazmanesh *et al.* 2020; Oliviero *et al.* 2020; Salazar de Pablo *et al.* 2020).

According to published data, COVID-19 has an initial incubation period that lasts between 2 and 19 days from exposure. Within that period, there is normally a symptomatic phase characterized by fever, chest discomfort, pulmonary and gastrointestinal manifestations. Finally, a COVID-19 infection may result in a complete recovery or, alternatively, may lead to death (Gulati *et al.* 2020; Khalili *et al.* 2020; Prajapati *et al.* 2020; Rajendran *et al.* 2020; Wu *et al.* 2020).

9.2.3. *Diagnosis*

Diagnosis of a COVID-19 case includes a number of criteria such as epidemiological characteristics, the underlying symptomatology of the probable patient and laboratory confirmation. Initially, personal contact with a confirmed COVID-19 case, as well as travel to a widely infected area 14 days prior to the onset of symptoms, were considered to be high risk factors. The combination of symptoms as well as their evolution are also taken into account in comparison to the case reports published thus far.

Nevertheless, the most reliable method to determine the probable existence of a COVID-19 infection is laboratory testing using oropharyngeal, nasopharyngeal specimens or blood samples. It should be mentioned, however, that in the case of a negative result, testing should be repeated within the proceeding days. Two laboratory procedures are used for the testing of COVID-19: molecular and serological tests. The first, called the reverse transcription polymerase chain reaction (RT-PCR), detects elements of the genetic material of the virus, whereas the second detects antigens or antibodies – in other words, specific proteins that highlight an immune response to the COVID-19 infection. It must be stressed that molecular testing, which is also considered to be the most accurate, can show whether there is an ongoing infection from SARS-CoV-2, whereas the serological tests are used to see whether an individual was previously infected (Fang and Meng 2020; Gupta *et al.* 2020; Phipps *et al.* 2020; Shyu *et al.* 2020). However, it must be noted that the accuracy of the latter might be problematic for various reasons. For example, if an individual had been infected in the past by any kind of coronavirus, a serological test will present positive COVID-19 results.

9.2.4. *Epidemiology and transmission of COVID-19*

The first outbreak of the SARS-CoV-2 virus occurred in Wuhan, China in December 2019. It is likely to have originated in bats, which are also well-known hosts of various coronaviruses (Hampton 2005; Li *et al.* 2005; Banerjee *et al.* 2019). In addition, genome analyses, protein sequences and phylogenetic analyses indicated a 96.2% genomic similarity between COVID-19 and Bat CoV RaTG13 viruses, highlighting, at the same time, various species (turtles, pangolins, etc.), as intermediate viral hosts (Liu *et al.* 2005; Lam *et al.* 2020; Wu *et al.* 2020; Zhou *et al.* 2020).

COVID-19 has spread throughout the world, yet for a number of countries, no cases have been reported, such as North Korea, Turkmenistan and the Solomon Islands. Since the public release of the COVID-19 genetic sequence on 12 January 2020, studies are focused on an understanding of its genome sequence, its ability to replicate and its probable vulnerabilities. Mutation might be one of the most prominent characteristics of a virus that secures its propagation. Even though the mutational extent of COVID-19 is not well established at the time of writing, it seems probable given it is the basis of its survival mechanism (Chan *et al.* 2020; Pachetti *et al.* 2020). Recent genome-wide studies show frequent mutations in residues of protein structures, suggesting a probable correlation to the virus's adaptability and transmission (Kaushal *et al.* 2020; Laha *et al.* 2020).

On January 14, 2020, the World Health Organization (WHO) confirmed human-to-human transmission of COVID-19 (Figure 9.1). It rapidly became clear that the significant characteristic of this coronavirus is its strong ability to transmit, mainly through respiratory droplets produced by the coughing or sneezing of an infected individual. These droplets can easily find their way into the lungs of the new host via inhalation. Contact transmission via a contagious object has also been reported as a potential route. Finally, there is aerosol transmission, the process by which respiratory droplets mix into the air, become aerosols and lead to an infection when inhaled in large quantities (Adhikari *et al.* 2020; Xu *et al.* 2020).

This particular characteristic is considered to be the main reason why COVID-19 went from initial outbreak to high risk pandemic so rapidly. A few weeks after the initial outbreak, China reported a total of 44 hospitalized patients who were suffering from pneumonia of unknown etiology, whereas it was not until January 13, 2020 that Thailand confirmed the existence of an individual infected by the already identified novel coronavirus. By the late January, infections in China signaled a significant outbreak with 7,736 cases and 170 deaths. The first "European" case was reported in mid-January, and by the end of that month, there were 7,818 cases worldwide, which led to the declaration of a "Public Health Emergency of International Concern" by the WHO. During the first weeks of March, the SARS-CoV-2 or COVID-19 virus was characterized as a "pandemic", with 88,586 cases and 3,050 deaths globally.

The basic reproduction number (R_0), in other words the average number of infections generated by one individual, plays a substantial role in quantifying transmission of the virus. According to research works, during the outbreak, China presented an average 3.28 R_0 value that ranged between 2.0 and 6.49 depending on the analyses and sample sizes (Chen and Wang 2020; Liu *et al.* 2020; Zhao *et al.* 2020). During the same period, European countries presented a 4.22 mean R_0 whereas Romania, Germany, the Netherlands and Spain showed the highest values (5.19–6.06) (Linka *et al.* 2020).

The sudden outbreak of COVID-19 and the extent of the deaths worldwide – that soon surpassed those of SARS-CoV – raised essential questions concerning the nature of the virus and the probable populations at high risk. Until now, older age and underlying diseases (such as cardiovascular diseases, chronic respiratory problems and diabetes), as well as obesity, are considered to be positively correlated to a rapid and severe evolution of the infection (Davies *et al.* 2020; Fang *et al.* 2020; Huang *et al.* 2020; Khalili *et al.* 2020). In addition, the biological sex of the individual has been identified as a key epidemiological factor in the case of COVID-19. In almost all of the published case studies worldwide, males do seem to be generally more susceptible to the virus with fatality numbers presenting a male-to-female ratio of 2:1 (Falahi and Kenarkoohi 2020; Garcia 2020; Jin *et al.* 2020; Ryan *et al.* 2020; Sun *et al.* 2020).

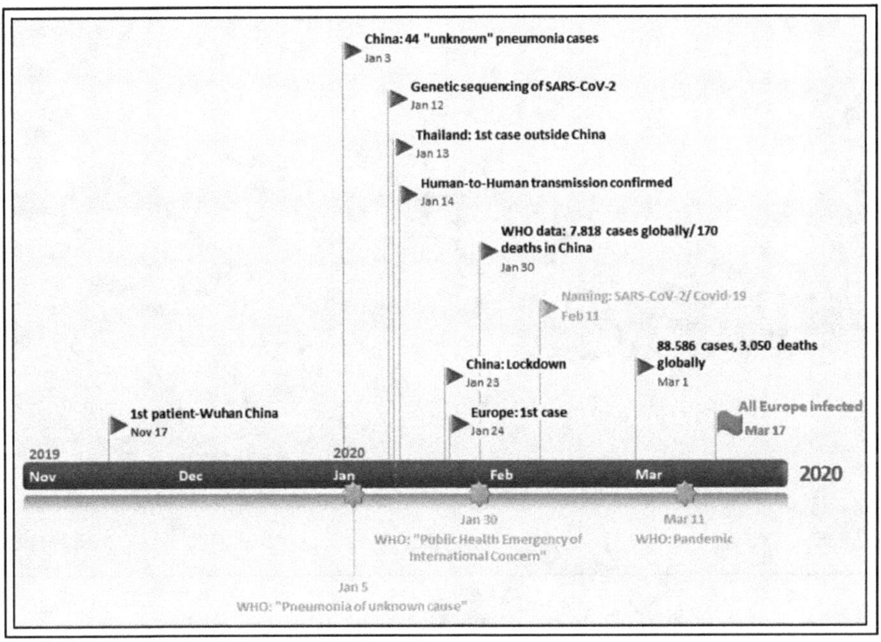

Figure 9.1. *COVID-19 pandemic timeline. For a color version of this figure, see www.iste.co.uk/dimotikalis/analysis2.zip*

9.2.5. *Country response measures*

The outbreak of the COVID-19 pandemic provoked a dynamic response from the medical community and the consequent global adoption of public measures, restrictions and surveillance strategies. The central aim of public measures was focused on containing the virus. Therefore, the initial focus globally was to shield the vulnerable populations and health workers, gaining, at oncc, the essential time needed to understand the disease and to build up therapeutic strategies.

Some of the most widely adopted measures included daily official (governmental and medical communities) communication to the public of current findings related to the pandemic, as well as guidance urging all citizens to self-monitor and to prevent interpersonal spread. In addition, there was the communication of protocols concerning hand hygiene, the use of masks, home disinfection, and the most up-to-date list of "suspicious" symptoms related to COVID-19, which preoccupied all forms of social media. The establishment of 24-hour health hotlines also became a global phenomenon.

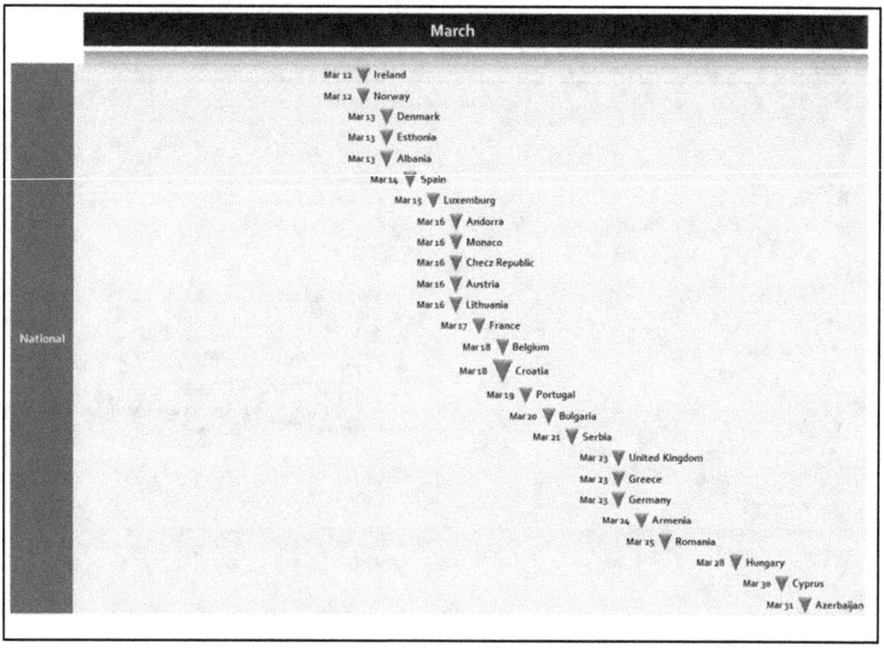

Figure 9.2. *European national lockdown timeline. For a color version of this figure, see www.iste.co.uk/dimotikalis/analysis2.zip*

Nevertheless, governmental policy responses varied, with some countries adopting radical measures and others maintaining a more "individualized" perspective. The reasons that provoked such variability are related to each governmental policy and strategy, to the level of competence of each national health care system and to the assumed extent of the COVID-19 outbreak in each country. In general, European countries presented a rather uniform response. From approximately mid-March, 25 countries implemented national lockdowns and 9 established regional lockdowns; this included social distancing, a ban on mass gatherings, shutdown of public institutions, border closures and travel bans.

Conversely, countries such as Sweden, Belarus, Moldova and Bosnia and Herzegovina followed a different strategy: there were generally no mandatory restrictions such as lockdowns, but social distancing rules, daily communication concerning the outbreak, and official advice on self-hygiene and quarantine of the infected (Figures 9.2 and 9.3).

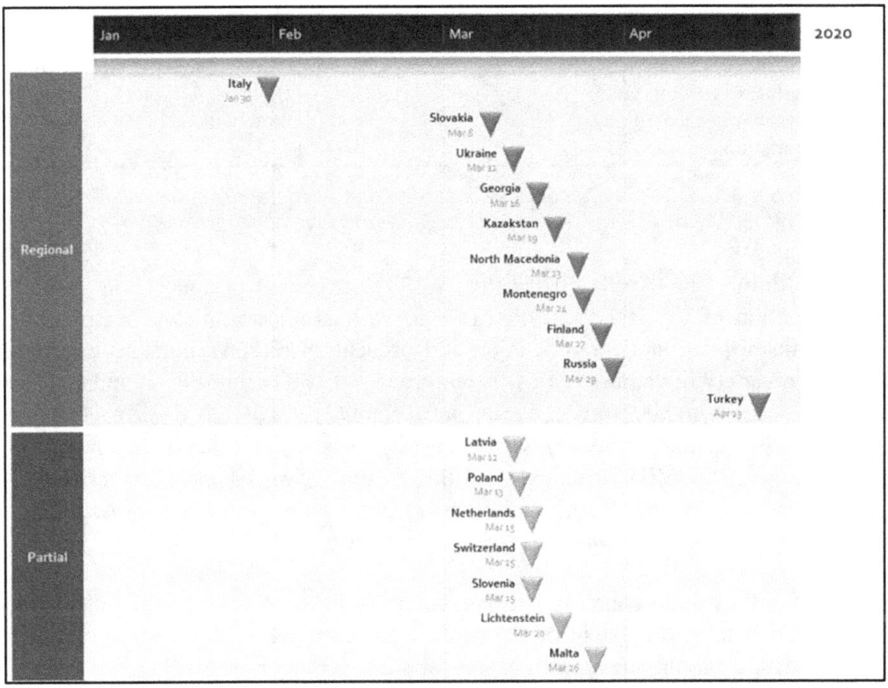

Figure 9.3. *European regional and partial lockdown timeline. For a color version of this figure, see www.iste.co.uk/dimotikalis/analysis2.zip*

Surveillance strategy plays an essential role in the global response to the COVID-19 outbreak. The lack of therapeutic or preventive means (i.e. vaccines, etc.) raises the need for a rigorous mapping of the potential of the COVID-19 infection and of its pattern of transmission. Soon after the European outbreak of COVID-19 in March, the World Health Organization hastened to provide the three-scale protocol in reporting cases (Table 9.1). However, it is notable that these protocols, though crucial in quantifying data, were not mandatory; instead their use relied on the willingness of each country.

Case no. 1	Case no. 2	Case no. 3
Patient with acute respiratory tract infection (sudden onset of at least one of the following: cough, fever, shortness of breath) **AND** with no other etiology to explain symptoms **AND** with a history of travel or residence in a country/area reporting local or community transmission during the 14 days prior to symptom onset	Patient with any acute respiratory illness **AND** having been in close contact with a confirmed or probable COVID-19 case in the last 14 days prior to onset of symptoms	Patient with severe acute respiratory infection (fever and at least one sign/symptom of respiratory disease e.g. cough, fever and shortness breath) **AND** requiring hospitalization **AND** with no other etiology that fully explains the clinical presentation

Table 9.1. *COVID-19 suspected case criteria (adapted from the WHO: https://www.who.int/emergencies/diseases/novelcoronavirus-2019)*

In addition, in April 2020, the WHO presented protocols on clinical documentation of COVID-19 deaths. Due to various uncertainties concerning the virus' transmission, and for protection and prevention reasons, diagnostic criteria included a variety of characteristics. Therefore, a "COVID-19 death" included: (1) *a confirmed case with no former recovery period* and (2) *a probable or suspected case with no other distinct cause of death*, but also suspected cases with clinical or epidemiological COVID-19 diagnosis followed by: (1) *an inconclusive test result*, (2) *a negative laboratory result* and even (3) *cases with chronic diseases, such as HIV, which should not be considered as the primary cause of death.*

Finally, there was variability between countries in terms of testing policies. This was related both to the extent of the applied procedures (i.e. the number of tested individuals per community) and of the applied protocols regarding the criteria according to which someone should be considered as a COVID-19 patient.

However, testing policy in general is linked to a plethora of agents. These vary significantly from country to country. Some of them are as follows:

1) Countries, as mentioned above, did not address the COVID-19 pandemic in a uniform way. For example, Sweden and the United Kingdom versus Germany (as well as other countries) are a characteristic example of how perceptions of the severity of a threat may vary.

2) Testing policy is directly related to the efficiency of each health care system, to its dynamic, completeness and to its coverage. It can be reasonably assumed that poorer and less developed countries faced more difficulties in the implementation of an effective testing policy, mainly due to infrastructure problems. In addition, social strata, socioeconomic and educational status of the employed populations, marginalization as well as social and cultural intra-population discrepancies are policy-related agents.

3) It has been observed that the economic status of a country played an essential role in the type of policy adopted in the face of the pandemic. This is not only related, for example, to the economic costs incurred by PCR tests or by medical/hospital treatments, but also to the fact that poorer countries are expected to have limited resources in consumables, antivirus drugs, up-to-date medical information and so on. Above all, the economic cost generated by the pandemic is mainly related to the degree of economic recession undergone in each of these countries.

If a person falls ill, do they have the same access to testing as in all the other European countries? Thus far, and according to the aforementioned issues, the answer appears to be negative.

9.2.6. *The role of statistical research in the case of COVID-19 and its challenges*

The importance of statistical research and epidemiological modeling is undeniable, especially in the case of viral outbreaks. Therefore, as has been shown in the case of COVID-19 and its unique high-speed spread, epidemiology has served as a primary means of information.

So far, analysis tools, indexes for the quantification of the disease's characteristics, forecasting models to observe the evolution of the virus, as well as epidemiological schemes that demonstrate transmission, contagiousness and vulnerability of populations and groups, have been used as a support to medical research and governmental strategies. However, despite this enormous contribution, the certainty and objectivity of generated results raise an important issue. As always, issues related to sample sizes and to the objectivity of the implemented parameters are severe limitations that might drastically alter all prognoses and outcomes.

For example, in the case of reproduction rates and transmission potential, discrepancies in testing and surveillance policies between countries, as well as the undefined number of asymptomatic patients are substantial limitations. In addition, the same applies for an understanding of the COVID-19 mortality dynamic. In such cases, the role of the employed protocols and criteria used for reporting cases and deaths, the wide range of clinical characteristics that are being reported as COVID-19 deaths, as well as the significant differences in global health systems and their policies, might produce high biases in statistical analyses and epidemiological models.

9.3. Materials and analyses

For our research, we collected data from all the daily updates from official online resources. World Health Organization (WHO) and Centers for Disease Control and Prevention (CDC) databases (https://www.who.int/), Worldmeter (https://www.worldometers.info/coro-navirus/), Our World in Data (https://ourworldindata.org) and European Commission statistics (https://ec.europa.eu), as well as countries' governmental websites, were used for the extraction of information concerning statistics, risk factors and prevention measurements.

According to the official data provided up until May 2020, we investigated the number of tests applied, as well as the reported COVID-19 cases and deaths for each of the European countries. A simple way to report such data is by presenting a time series of the absolute number of cases observed in each country by gender and applying a mathematical model to it. However, such an illustration – while very useful for recording registered cases (hospitalized or not) on a daily basis and for the development of appropriate patient institutionalization policies – is of little importance for the comprehension of the prevalence of the SARS-CoV-2 virus and the COVID-19 disease in each population and for intra/inter-population comparisons. The reason is simple: different countries include different populations. One solution could be the calculation of rates, i.e. cases or deaths per 100,000 or 1,000,000 population per gender.

Unfortunately, this solution is also problematic. Each population presents different age clusters: to put it simply, young children, juveniles, young adults and all the other ages until the last stages of the human lifecycle. Does the virus affect people of all ages? In addition, is the probability of being a COVID-19 patient or virus carrier the same in all age groups? What is the probability if you are a carrier/patient of being tested for COVID-19, not forgetting at the same time the variability in the applied testing policies?

The last question is the most important concerning the actual prevalence of COVID-19 within a specific population. It is well known that most younger people are asymptomatic during the carrier stage. Then reasonably enough, these people will never be examined for this disease even though they can transmit it to other people. The same of course holds for people of other ages being either asymptomatic carriers or having mild symptoms. In the latter case, it depends on the country if one is to be tested for the coronavirus. Therefore, it is expected that most of the asymptomatic carriers in a country would be unknown and the same occurs with many patients who have mild symptoms.

From this point of view, it is actually difficult and complex to estimate the prevalence of the present virus, whereas all available numbers correspond only to the individuals that have been tested. Is it possible to consider these people as a random sample in order to refer to the real prevalence of the virus within the population and thus of the disease itself? The answer is not simple since all of the COVID-19 carriers are not equally included within these samples.

A solution to the above could be the development of stratified random sampling testing procedures among populations. However, such a solution is very costly, time-consuming and needs to be repeated periodically in a dynamic environment as it is a pandemic outbreak. A further complication which needs to be addressed emerges from the fact that some carriers could have a very low "viral load" at the time of the sampling. In turn, they could initially get a false negative result, whereas a repeat test a few days later would be COVID-19 positive.

To conclude, it is questionable whether the available data can describe the real prevalence of COVID-19. Even though such data could be available, for inter- and intra-population comparisons, the results of the analysis require a disregard of any differentiation in social and population structures. It is self-evident that a population mainly consisting of aged individuals would respectively present higher numbers of COVID-19 deaths or elevated COVID-19 cases than others, a well-known phenomenon in demography.

In this way, the following section presents some of the preliminary statistical analyses.

9.4. The first phase of the pandemic

Statistical analyses indicate that the average number of total COVID-19 cases per 100,000 population and country is 220.2 ± 303.8. Until May 2020, the largest percentage of the employed countries reported more than 100 cases per 100,000 individuals. The classification of the European countries on the basis of the standard deviation of the above-mentioned mean "prevalence rate" revealed a four-scale scheme (Figures 9.4a, b and c): 25 countries belong to a differentiated group in which the number of cases per 100,00 population is low (-0.50 Std. Dev. group), 19 lie in the range -0.50 to 0.50 standard deviations, whereas only nine lie in the range between 0.50 and 1.50. In a small number of countries, the prevalence of the disease was more intense.

As shown in Figure 9.4a, the low prevalence countries are of Eastern Europe and the Balkans. Because many of these countries are among the poorer in Europe, an open question remains to be addressed. Does this finding represent a smaller or delayed development of the pandemic, or might it also be attributed to a partial or total inability to identify the intensity of the pandemic in some of these countries? Obviously enough, a lower prevalence rate characterizes this part of the continent, although health infrastructure, economic and other types of problems may have played an important role in the quantification of the pandemic. This question remains open for further investigation.

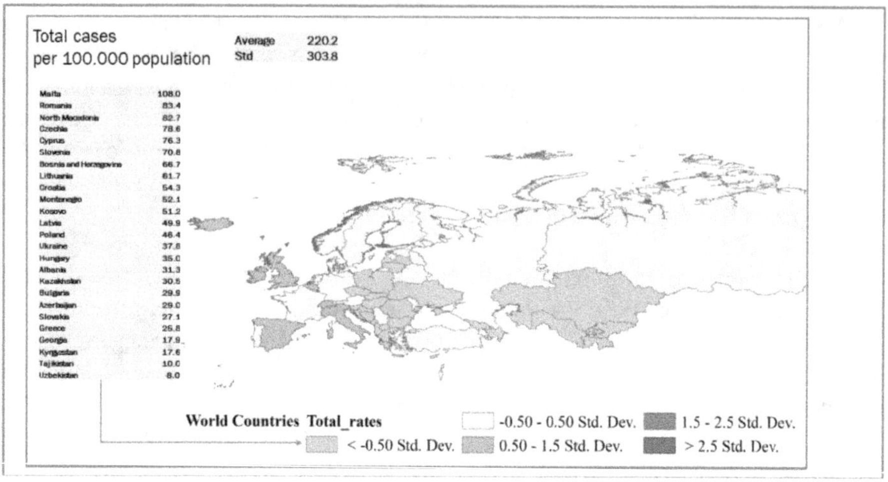

Figure 9.4a. *Total cases per 100,000 population with low prevalence. For a color version of this figure, see www.iste.co.uk/dimotikalis/analysis2.zip*

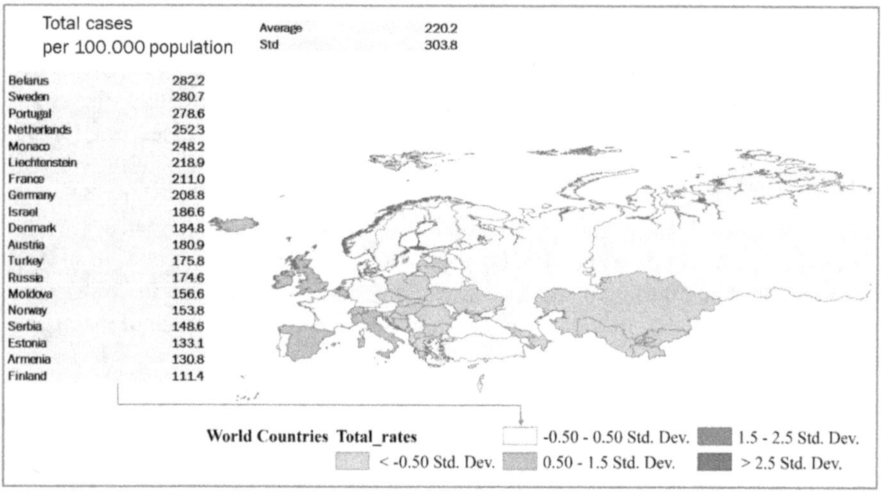

Figure 9.4b. *Total cases per 100,000 population with medium prevalence. For a color version of this figure, see www.iste.co.uk/dimotikalis/analysis2.zip*

Many countries lie very close to the European average (Figure 9.4b). These are spatially, politically, economically and culturally diversified, spread all over the continent. Undoubtedly, some of the reservations expressed in the previous paragraph also apply to this group. The group of high prevalence countries consists of Iceland, Spain, Ireland, Belgium, the Isle of Man, the Faroe Islands, Italy, Switzerland and the United Kingdom (Figure 9.4c). To them, Andorra, Luxembourg and San Marino might be added, although their population is limited and such an estimation relies on a limited number of cases.

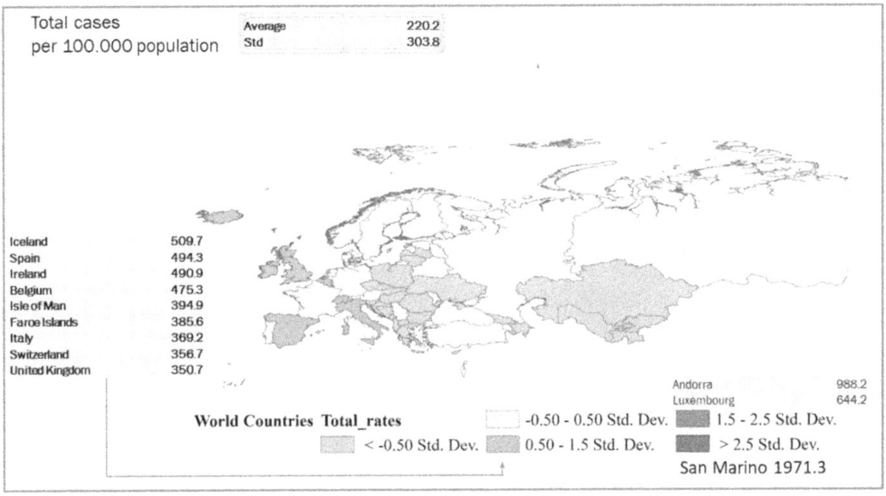

Figure 9.4c. *Total cases per 100,000 population with a high prevalence. For a color version of this figure, see www.iste.co.uk/dimotikalis/analysis2.zip*

However, we could expect a positive relationship between the COVID-19 detection test rate and reported cases of the disease if both the prevalence of the disease and the testing policy were uniformly distributed among the countries. However, even though such a linear relationship does exist, as shown in Figure 9.5, the coefficient of determination of the fit is only 34%. This low coefficient indicates the diversity of the test rate and the relevant testing capabilities and policies among the European countries, in relation to the reported COVID-19 cases. For example, in Israel (IS), the test rate is the highest in comparison to all the other countries, despite the fact that this country is classified in the medium prevalence group. Also, Lithuania (LT), a country of the low prevalence group, had the third highest test rate in comparison to all the other countries. On the contrary, Italy (IT), of the high prevalence group, had almost the same test rate as Latvia (LV), of the low prevalence group. Finally, in many countries (Greece, Bulgaria, Ukraine, Hungary, Poland, Turkey, Russia, Finland, Czech Republic, Slovenia, Slovakia), both test and case rates were low.

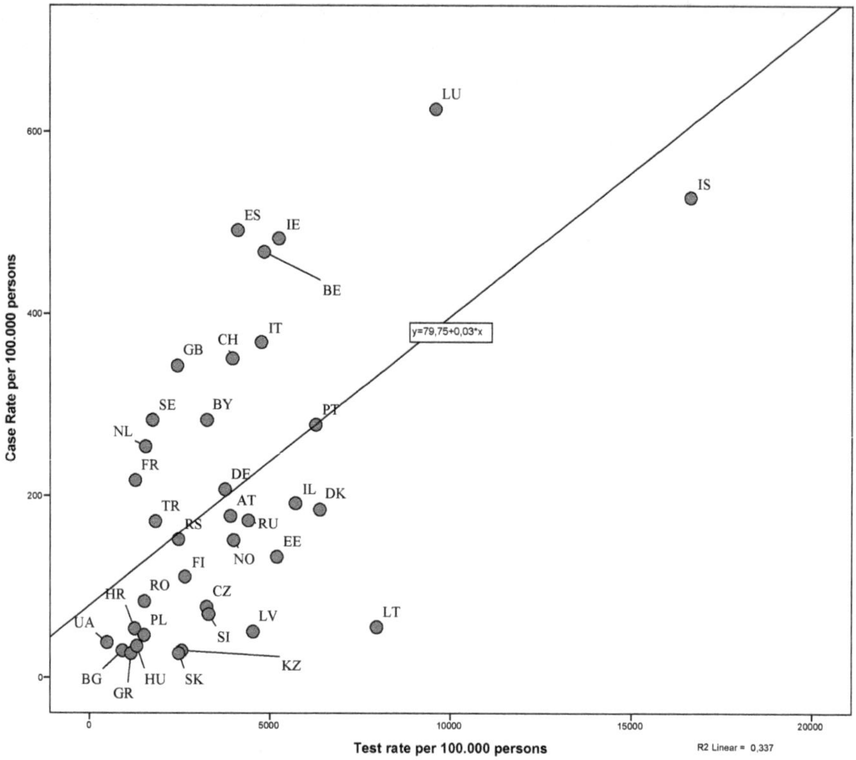

Figure 9.5. *Tests and cases per 100,000 persons. For a color version of this figure, see www.iste.co.uk/dimotikalis/analysis2.zip*

As a result of the previous findings, there is no linear relationship between the test rate and the number of cases recorded per thousand tests and country (Figure 9.6). In this illustration, it is seen that in the most discrete position lie France (FR), Sweden (SE), the Netherlands (NL), Great Britain (GB) and Spain (ES), i.e. group A in Figure 9.6. In this, more patients were recorded per performed tests; thus, the countries were more frugal in conducting patient-related testing. The second group, B, consists of Ukraine (UA), Turkey (TR), Belarus (BY), Switzerland (CH), Belgium (BE), Ireland (IE) and Italy (IT), and lies in an intermediate position. The last group (group C) consists of all the other countries, it is rather heterogeneous, and fewer patients were recorded per tests performed. An obvious conclusion then is that the observed heterogeneity serves negatively any effort for inter-country comparison: the different COVID-19 testing policies most likely led to possible patient recording leaks in many countries and thus any available data is not directly comparable.

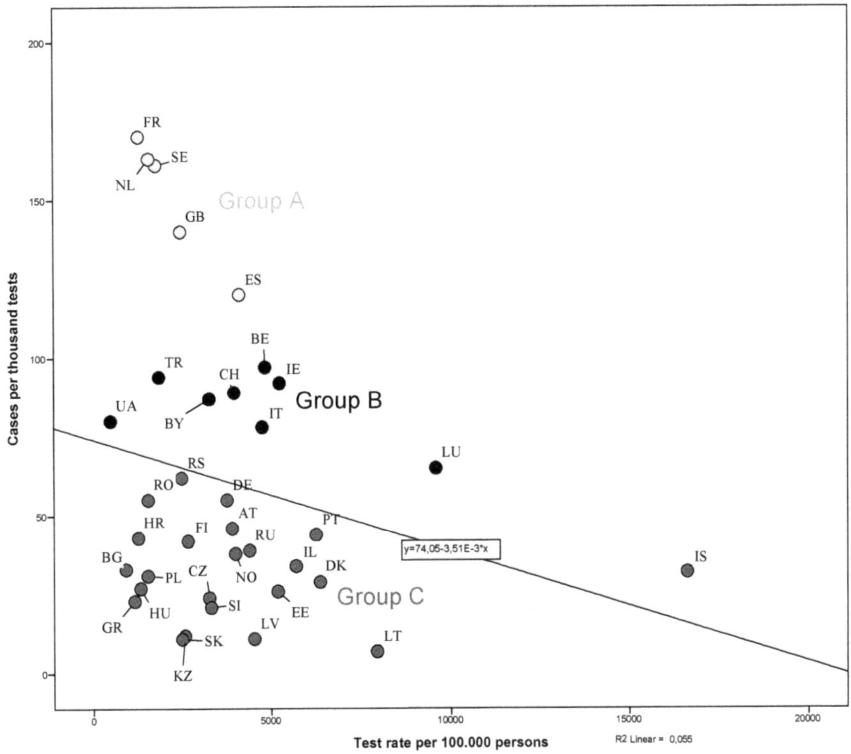

Figure 9.6. *Cases per 1,000 tests. For a color version of this figure, see www.iste.co.uk/dimotikalis/analysis2.zip*

In any case, Figure 9.7 illustrates the number of deaths per 100,000 positive COVID-19 recorded cases. However, it must be stressed that this is not a measure of mortality. Such an estimation requires age standardization of mortality rates in order to become comparable between the countries. Also, it requires detailed death by age data, which are either partially known or absent, as well as detailed information of the age structure of each population for the year 2020, which is unavailable for the moment. It is also not a measure of fatality. Such a measure requires detailed, accurate data of COVID-19 prevalence, which is largely problematic too, as discussed previously. Thus, the result presented in Figure 9.9 must be understood as a crude and limited estimation of the pandemic.

Keeping in mind the above, the European countries were classified in Figure 9.7 on the basis of standard deviation of the mean rate. The most diverse country is France, where the estimations show the vast effect of the COVID-19 pandemic.

However, it is important to bear in mind that France was also one of the most thrifty countries if one takes into consideration the relationship between the tests conducted and the COVID-19 recorded cases. The second group of countries is formed by Belgium, the United Kingdom, Italy, Hungary, the Netherlands and Sweden. Spain ranks in eighth position. All the other countries cluster into two groups. They are either very close to the European average (22 countries) or lower (22 countries).

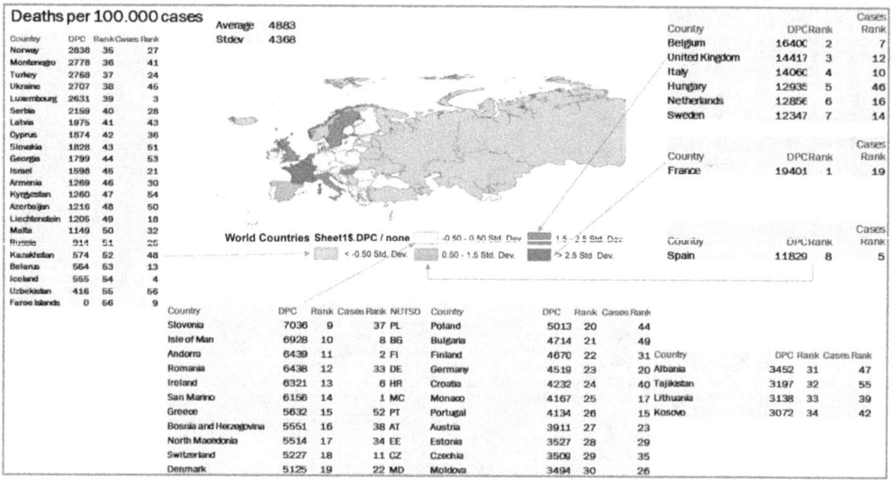

Figure 9.7. *Deaths per 100,000 cases. For a color version of this figure, see www.iste.co.uk/dimotikalis/analysis2.zip*

Thus, the COVID-19 pandemic is evident but diverse among the countries of the European continent, given the problems of the data, of course.

9.5. Concluding remarks

It is undeniable the fact that the COVID-19 pandemic is a significant ongoing menace to current times. The biological and epidemiological data we currently possess confirm that SARS-CoV-2 presents an elevated potential to mutate and transmit. However, even though the adoption of analytical models and generalizations concerning the process of the pandemic are very important, they involve various problems. Transnational discrepancies in the applied policies, surveillance and sampling strategies, differences in health systems and socioeconomic status lead to controversial statistics that need to be taken into consideration. The present study focuses on the first phase of the pandemic. A few months later, and after the raising of lockdowns, there seems to be a global resurgence in terms of COVID-19 cases. The forthcoming analyses, along with the results we will obtain by the promising vaccinations, will surely broaden our knowledge on this pandemic, offering at the same time new insights concerning the quantification of large-scale epidemiological data.

9.6. References

Abdin, S.M., Elgendy, S.M., Alyammahi, S.K., Alhamad, D.W., Omar, H.A. (2020). Tackling the cytokine storm in COVID-19, challenges, and hopes. *Life Sciences*, 257, 118054.

Abebe, E.C., Dejenie, T.A., Shiferaw, M.Y., Malik, T. (2020). The newly emerged COVID-19 disease: A systemic review. *Virology Journal*, 17(1), 96.

Adhikari, S.P., Meng, S., Wu, Y.J., Mao, Y.P., Ye, R.X., Wang, Q.Z., Sun, C., Sylvia, S., Rozelle, S., Raat, H., Zhou, H. (2020). Epidemiology, causes, clinical manifestation and diagnosis, prevention and control of coronavirus disease (COVID-19) during the early outbreak period: A scoping review. *Infectious Diseases of Poverty*, 9(1), 29.

Agarwal, S. and Agarwal, S.K. (2020). Endocrine changes in SARS-CoV-2 patients and lessons from SARS-CoV. *Postgraduate Medical Journal*, 96(1137), 412–416.

Aghagoli, G., Gallo Marin, B., Katchur, N.J., Chaves-Sell, F., Asaad, W.F., Murphy, S.A. (2020). Neurological involvement in COVID-19 and potential mechanisms: A review. *Neurocritical Care*, July 13, 1–10.

Banerjee, A., Kulcsar, K., Misra, V., Frieman, M., Mossman, K. (2019). Bats and coronaviruses. *Viruses*, 11(1), E41.

Bi, Q., Wu, Y., Mei, S., Ye, C., Zhou, X., Zhang, Z. (2020). Epidemiology and transmission of Covid-19 in 391 cases and 1286 of their close contacts in Shenzhen, China: A retrospective cohort study. *The Lancet Infectious Diseases*, 20(8).

Callaway, E., Cyranoski, D., Mallapaty, S., Stoye, E., Tollefsom, J. (2020). Coronavirus by the numbers. *Nature*, 579, 482–483.

Chan, J.F., Kok, K., Zhu, Z., Chu, H., To K.K., Yuan, S., Yuen, K.Y. (2020a). Genomic characterisation of the 2019 novel human-pathogenic coronavirus isolated from a patient with atypical pneumonia after visiting Wuhan. *Emerging Microbes & Infections*, 9, 221–236.

Chan, A.P., Choi, Y., Schork, N.J. (2020b). Conserved genomic terminals of SARS-CoV-2 as co-evolving functional elements and potential therapeutic targets. *BioRxiv: The Preprint Server for Biology*, 7(6), 190–207.

Dasari, C.M. and Bhukya, R. (2020). Comparative analysis of protein synthesis rate in COVID-19 with other human coronaviruses. *Infection, Genetics and Evolution: Journal of Molecular Epidemiology and Evolutionary Genetics in Infectious Diseases*, 85, 104–432.

Davies, N.G., Klepac, P., Liu, Y., Prem, K., Jit, M. (2020). CMMID COVID-19 working group, Eggo, R.M. Age-dependent effects in the transmission and control of COVID-19 epidemics. *Nature Medicine*, 26, 1205–1211.

Demirci Otluoglu, G., Yener, U., Demir, M.K., Yilmaz, B. (2020). Encephalomyelitis associated with Covid-19 infection: Case report. *British Journal of Neurosurgery*, 1–3.

European Centre for Disease Prevention and Control (2020). Clinical characteristics of COVID-19 [Online]. Available at: https://www.ecdc.europa.eu/en/covid-19/latest-evidence/clinical.

Falahi, S. and Kenarkoohi, A. (2020). Sex and gender differences in the outcome of patients with COVID-19. *Journal of Medical Virology*, 93(1), 151–152.

Fang, B. and Meng, Q.H. (2020). The laboratory's role in combating COVID-19. *Critical Reviews in Clinical Laboratory Sciences*, 1–15.

Fang, X., Li, S., Yu, H., Wang, P., Zhang, Y., Chen, Z., Li, Y., Cheng, L., Li, W., Jia, H., Ma, X. (2020). Epidemiological, comorbidity factors with severity and prognosis of COVID-19: A systematic review and meta-analysis. *Aging*, 12(13), 12493–12503.

Fiani, B., Covarrubias, C., Desai, A., Sekhon, M., Jarrah, R. (2020). A contemporary review of neurological sequelae of COVID-19. *Frontiers in Neurology*, 11, 640.

Garcia, L.P. (2020). Sex, gender and race dimensions in COVID-19 research. Dimensões de sexo, gênero e raça na pesquisa sobre COVID-19. *Epidemiologia e servicos de saude: revista do sistema unico de saude do brasil*, 29(3), e20202207.

Guan, W.J., Ni, Z.Y., Hu, Y. (2020). Clinical characteristics of coronavirus disease 2019 in China. *New England Journal of Medicine*, 382, 1708–1720.

Gupta, A., Madhavan, M.V., Sehgal, K., Nair, N., Mahajan, S., Sehrawat, T.S., Bikdeli, B., Ahluwalia, N., Ausiello, J.C., Wan, E.Y., Freedberg, D.E., Kirtane, A.J., Parikh, S.A., Maurer, M.S., Nordvig, A.S., Accili, D., Bathon, J.M., Mohan, S., Bauer, K.A., Leon, M.B., Landry, D.W. (2020). Extrapulmonary manifestations of COVID-19. *Nature Medicine*, 26(7), 1017–1032.

Gulati, A., Pomeranz, C., Qamar, Z., Thomas, S., Frisch, D., George, G., Summer, R., De Simone, J., Sundaram, B. (2020). A comprehensive review of manifestations of novel coronaviruses in the context of deadly COVID-19 global pandemic. *The American Journal of the Medical Sciences*, 360(1), 5–34.

Hampton, T. (2005). Bats may be SARS reservoir. *JAMA*. 294(18), 2291.

Hardenberg, J.H. and Luft, F.C. (2020). Covid-19, ACE2, and the kidney. *Acta Physiologica (Oxford, England)*, 230(1), e13539.

Hoffmann, M., Kleine-Weber, H., Krüger, N., Müeller, M.A., Drosten, C., Pöhlmann, S. (2020). The novel coronavirus 2019 (2019-nCoV) uses the SARS-coronavirus receptor ACE2 and the cellular protease TMPRSS2 for entry into target cells [Online]. Available at: https://doi.org/10.1101/2020.01.31.929042.

Honigsbaum, M. (2020). Revisiting the 1957 and 1968 influenza pandemics. *Lancet (London, England)*, 395(10240), 1824–1826.

Huang, C., Wang, Y., Li, X. (2020). Clinical features of patients infected with 2019 novel coronavirus in Wuhan, China. *Lancet*, 395(10223), 497– 506.

Jang, Y. and Seo, S.H. (2020). Gene expression pattern differences in primary human pulmonary epithelial cells infected with MERS-CoV or SARS-CoV-2. *Archives of Virology*, 165, 2205–2211.

Jin, J.M., Bai, P., He, W., Wu, F., Liu, X.F., Han, D.M., Liu, S., Yang, J.K. (2020). Gender differences in patients with COVID-19: Focus on severity and mortality. *Frontiers in Public Health*, 8, 152.

Kaswa, R. and Govender, I. (2020). Novel coronavirus pandemic: A clinical overview. *South African Family Practice (2004)*, 62(1), e1–e5.

Kaushal, N., Gupta, Y., Goyal, M., Khaiboullina, S.F., Baranwal, M., Verma, S.C. (2020). Mutational frequencies of SARS-CoV-2 genome during the beginning months of the outbreak in U.S.A. *Pathogens (Basel, Switzerland)*, 9(7), E565.

Khalili, M., Karamouzian, M., Nasiri, N., Javadi, S., Mirzazadeh, A., Sharifi, H. (2020). Epidemiological characteristics of COVID-19: A systematic review and meta-analysis. *Epidemiology and Infection*, 148, e130.

Kilbourne, E.D. (2006). Influenza pandemics of the 20th century. *Emerging Infectious Diseases*, 12(1), 9–14.

Kreitmann, L., Monard, C., Dauwalder, O., Simon, M., Argaud, L. (2020). Early bacterial co-infection in ARDS related to COVID-19. *Intensive Care Medicine*, July 13, 1–3.

Laha, S., Chakraborty, J., Das, S., Manna, S.K., Biswas, S., Chatterjee, R. (2020). Characterisations of SARS-CoV-2 mutational profile, spike protein stability and viral transmission. *Infection, Genetics and Evolution: Journal of Molecular Epidemiology and Evolutionary Genetics in Infectious Diseases*, 85, 104445.

Lai, M.M.C. and Holmes, K.V. (2001). The viruses and their replication. In *Coronaviridae*, 4th edition, Knipe, D.M. and Howley, P.M. (eds). Lippincott, Williams & Wilkins, Philadelphia.

Lam, T.T., Jia, N., Zhang, Y. (2020). Identifying SARS-CoV-2-related coronaviruses in Malayan pangolins. *Nature*, 583, 282–285.

Li, F. (2016). Structure, function, and evolution of coronavirus spike proteins. *Annual Review of Virology*, 3, 237–261.

Li, W., Shi, Z., Yu, M., Ren, W., Smith, C., Epstein, J.H. (2005). Bats are natural reservoirs of SARS-like coronaviruses. *Science*, 310(5748), 676–679.

Liao, D., Zhou, F., Luo, L., Xu, M., Wang, H., Xia, J., Gao, Y., Cai, L., Wang, Z., Yin, P., Wang, Y., Tang, L., Deng, J., Mei, H., Hu, Y. (2020). Haematological characteristics and risk factors in the classification and prognosis evaluation of COVID-19: A retrospective cohort study. *The Lancet. Haematology*, S2352-3026(20)30217-9.

Linka, K., Peirlinck, M., Kuhl, E. (2020). The reproduction number of COVID-19 and its correlation with public health interventions. *Computational Mechanics*, July 28, 1–16.

Liu, J.W., Bi, Y., Wang, D., Gao, G.F. (2018). On the centenary of the Spanish flu: Being prepared for the next pandemic. *Virologica Sinica*, 33, 463–466.

Liu, Z., Xiao, X., Wei, X., Li, J., Yang, J., Tan, H. (2020a). Composition and divergence of coronavirus spike proteins and host ACE2 receptors predict potential intermediate hosts of SARS-CoV-2. *J. Med. Virol.*, 92(6),595-601.

Liu, Y., Gayle, A.A., Wilder-Smith, A., Rocklöv, J. (2020b). The reproductive number of COVID-19 is higher compared to SARS coronavirus. *Journal of Travel Medicine*, 27(2), taaa021.

Lu, R., Zhao, X., Li, J., Niu, P., Yang, B., Wu, H., Wang, W., Song, H., Huang, B., Zhu, N., Bi, Y., Ma, X., Zhan, F., Wang, L., Hu, T., Zhou, H., Hu, Z., Zhou, W., Zhao, L., Chen, J., Meng, Y., Wang, J., Lin, Y., Yuan, J., Xie, Z., Ma, J., Liu, W.J., Wang, D., Xu, W., Holmes, E.C., Gao, G.F., Wu, G., Chen, W., Shi, W., Tan, W. (2020). Genomic characterisation and epidemiology of 2019 novel coronavirus: Implications for virus origins and receptor binding. *Lancet*, 395(10224), 565–574.

Malik, Y.S., Sircar, S., Bhat, S., Sharun, K., Dhama, K., Dadar, M., Tiwari, R., Chaicumpa, W. (2020). Emerging novel coronavirus (2019-nCoV)-current scenario, evolutionary perspective based on genome analysis and recent developments. *The Veterinary Quarterly*, 40(1), 68–76.

Marra, M.A., Jones, S.J., Astell, C.R., Holt, R.A., Brooks-Wilson, A., Butterfield, Y.S., Khattra, J., Asano, J.K., Barber, S.A., Chan, S.Y., Cloutier, A., Coughlin, S.M. (2003). The genome sequence of the SARS-associated coronavirus. *Science*, 1399–1404.

Masters, S.P. (2006). The molecular biology of coronaviruses. *Advances in Virus Research*, 66, 193–292.

Momtazmanesh, S., Shobeiri, P., Hanaei, S., Mahmoud-Elsayed, H., Dalvi, B., Malakan Rad, E. (2020). Cardiovascular disease in COVID-19: A systematic review and meta-analysis of 10,898 patients and proposal of a triage risk stratification tool. *The Egyptian Heart Journal: (E.H.J.): Official Bulletin of the Egyptian Society of Cardiology*, 72(1), 41.

Nepal, G., Rehrig, J.H., Shrestha, G.S., Shing, Y.K., Yadav, J.K., Ojha, R., Pokhrel, G., Tu, Z.L., Huang, D.Y. (2020). Neurological manifestations of COVID-19: A systematic review. *Critical Care (London, England)*, 24(1), 421.

Neuman, B.W., Kiss, G., Kunding, A.H., Bhella, D., Baksh, M.F., Connelly, S., Droese, B., Klaus, J.P., Makino, S., Sawicki, S.G. (2011). A structural analysis of m protein in coronavirus assembly and morphology. *J. Struct. Biol.*, 174(1), 11–22.

Oliviero, A., de Castro, F., Coperchini, F., Chiovato, L., Rotondi, M. (2020). COVID-19 pulmonary and olfactory dysfunctions: Is the chemokine CXCL10 the common denominator? *The Neuroscientist: A Review Journal Bringing Neurobiology, Neurology and Psychiatry*, July 13, 1073858420939033.

Pachetti, M., Marini, B., Benedetti, F., Giudici, F., Mauro, E., Storici, P., Masciovecchio, C., Angeletti, S., Ciccozzi, M., Gallo, R.C., Zella, D., Ippodrino, R. (2020). Emerging SARS-CoV-2 mutation hot spots include a novel RNA-dependent-RNA polymerase variant. *Journal of Translational Medicine*, 18(1), 179.

Pan, L., Mu, M., Yang, P., (2020). Clinical characteristics of COVID-19 patients with digestive symptoms in Hubei, China: A descriptive, crosssectional, multicenter study. *American Journal of Gastroenterology*, 115(5), 766–773.

Petersen, E., Koopmans, M., Go, U., Hamer, D.H., Petrosillo, N., Castelli, F., Storgaard, M., Al Khalili, S., Simonsen, L. (2020). Comparing SARS-CoV-2 with SARS-CoV and influenza pandemics. *The Lancet. Infectious Diseases*, S1473-3099(20)30484-9.

Phipps, W.S., SoRelle, J.A., Li, Q.Z., Mahimainathan, L., Araj, E., Markantonis, J., Lacelle, C., Balani, J., Parikh, H., Solow, E.B., Karp, D.R., Sarode, R., Muthukumar, A. (2020). SARS-CoV-2 antibody responses do not predict COVID-19 disease severity. *American Journal of Clinical Pathology*, 154(4), 459–465.

Poyiadji, N., Shahin, G., Noujaim, D., Stone, M., Patel, S., Griffith, B. (2020). COVID-19-associated acute hemorrhagic necrotising encephalopathy: C.T. and M.R.I. features [Online]. Available at: https://doi.org/10.1148/radiol.2020201187.

Prajapati, S., Sharma, M., Kumar, A., Gupta, P., Narasimha Kumar, G.V. (2020). An update on novel COVID-19 pandemic: A battle between humans and virus. *European Review for Medical and Pharmacological Sciences*, 24(10), 5819–5829.

Rajendran, D.K., Rajagopal, V., Alagumanian, S., Santhosh Kumar, T., Sathiya Prabhakaran, S.P., Kasilingam, D. (2020). Systematic literature review on novel corona virus SARS-CoV-2: A threat to human era. *Virusdisease*, 31(2), 161–173.

Rota, P.A., Oberste, M.S., Monroe, S.S., Nix, W.A., Campagnoli, R., Icenogle, J.P., Penaranda, S., Bankamp, B., Maher, K., Chen, M.H., Tong, S., Tamin, A. (2003). Characterization of a novel coronavirus associated with severe acute respiratory syndrome. *Science*, 1394–1399.

Ruch, T.R. and Machamer, C.E. (2012). The coronavirus e protein: Assembly and beyond. *Viruses*, 4(3), 363–382.

Ryan, N.E. and El Ayadi, A.M. (2020). A call for a gender-responsive, intersectional approach to address COVID-19. *Global Public Health*, 1–9.

Salazar de Pablo, G., Vaquerizo-Serrano, J., Catalan, A., Arango, C., Moreno, C., Ferre, F., Shin, J.I., Sullivan, S., Brondino, N., Solmi, M., Fusar-Poli, P. (2020). Impact of coronavirus syndromes on physical and mental health of health care workers: Systematic review and meta-analysis. *Journal of Affective Disorders*, 275, 48–57.

Schoeman, D. and Fielding, B.C. (2019). Coronavirus envelope protein: Current knowledge. *Virology Journal*, 16(1), 69.

Schwartz J.L. (2018). The spanish flu, epidemics, and the turn to biomedical responses. *American Journal of Public Health*, 108(11), 1455–1458.

Shereen, M.A., Khan, S., Kazmi, A., Bashir, N., Siddique, R. (2020). COVID-19 infection: Origin, transmission, and characteristics of human coronaviruses. *Journal of Advanced Research*, 24, 91–98.

Shyu, D., Dorroh, J., Holtmeyer, C., Ritter, D., Upendran, A., Kannan, R., Dandachi, D., Rojas-Moreno, C., Whitt, S.P., Regunath, H. (2020). Laboratory tests for COVID-19: A review of peer-reviewed publications and implications for clinical ulse. *Missouri Medicine*, 117(3), 184–195.

Spiteri, G., Fielding, J., Diercke, M., Campese, C., Enouf, V., Gaymard, A., Bella, A., Sognamiglio, P., Sierra Moros, M.J., Riutort, A.N., Demina, Y.V., Mahieu, R., Broas, M., Bengnér, M., Buda, S., Schilling, J., Filleul, L., Lepoutre, A., Saura, C., Mailles, A., Ciancio, B.C. (2020). First cases of coronavirus disease 2019 (COVID-19) in the WHO European region, 24 January to 21 February 2020. *Euro Surveillance: Bulletin Europeen sur les maladies transmissibles = European communicable disease bulletin*, 25(9), 2000178.

Sui, J., Deming, M., Rockx, B., Liddington, R.C., Zhu, Q.K., Baric, R.S., Marasco, W.A. (2014). Effects of human anti-spike protein receptor binding domain antibodies on severe acute respiratory syndrome coronavirus neutralisation escape and fitness. *Journal of Virology*, 88(23), 13769–13780.

Tabata, S., Imai, K., Kawano, S., Ikeda, M., Kodama, T., Miyoshi, K., Obinata, H., Mimura, S., Kodera, T., Kitagaki, M., Sato, M., Suzuki, S., Ito, T., Uwabe, Y., Tamura, K. (2020). Clinical characteristics of COVID-19 in 104 people with SARS-CoV-2 infection on the diamond princess cruise ship: A retrospective analysis. *The Lancet. Infectious Diseases*, S1473-3099(20)30482-5.

Tort, F.L., Castells, M., Cristina, J. (2020). A comprehensive analysis of genome composition and codon usage patterns of emerging coronaviruses. *Virus Research*, 283, 197976.

Wang, D., Hu, B., Hu, C. (2020a). Clinical characteristics of 138 hospitalised patients with 2019 novel coronavirus-infected pneumonia in Wuhan, China. *JAMA*, 323(11), 1061–1069.

Wang, W., Chen, Y., Wang, Q., Cai, P., He, Y., Hu, S. (2020b). The transmission dynamics of SARSCOV-2 in China: Modeling study and the impact of public health interventions [Online]. Available at: 10.1101/2020.03.24.20036285.

Wege, H., Siddell, S., ter Meulen, V. (1982). The biology and pathogenesis of coronaviruses. In *Current Topics in Microbiology and Immunology*, Cooper, M., Henle, W., Hofschneider, P.H., Koprowski, H., Melchers, F., Rott, R., Schweiger, H.G., Vogt, P.K., Zinkernagel, R. (eds). Springer, Berlin, Heidelberg.

Weiss, S.R. and Navas-Martin, S. (2005). Coronavirus pathogenesis and the emerging pathogen severe acute respiratory syndrome coronavirus. *American Society for Microbiology Journals*, 69(4), 635–664.

Wu, C. (2020). Risk factors associated with acute respiratory distress syndrome and death in patients with coronavirus disease 2019 pneumonia in Wuhan, China. *JAMA Internal Medicine*, 180(7), 934–943.

Wu, F., Zhao, S., Yu, B., Chen, Y.M., Wang, W., Song, Z.G., (2020). A new coronavirus associated with human respiratory disease in China. *Nature*, 579, 265–269.

Xu, X., Chen, P., Wang, J., Feng, J., Zhou, H., Li, X, (2020). Evolution of the novel coronavirus from the ongoing Wuhan outbreak and modeling of its spike protein for risk of human transmission. *Science China Life Sciences*, 63, 457–460.

Ye, Q., Wang, B., Zhang, T., Xu, J., Shang, S. (2020). The mechanism and treatment of gastrointestinal symptoms in patients with COVID-19. *American Journal of Physiology. Gastrointestinal and Liver Physiology*, 319(2), G245-G252.

Zaki, A.M., van Boheemen, S., Bestebroer, T.M., Osterhaus, A.D., Fouchier, R.A. (2012). Isolation of a novel coronavirus from a man with pneumonia in Saudi Arabia. *New England Journal of Medicine*, 8, 1814–1820.

Zhang, W., Zhao, Y., Zhang, F., (2020). The use of anti-inflammatory drugs in the treatment of people with severe coronavirus disease 2019 (COVID19): The experience of clinical immunologists from China. *Clinical Immunology*, 214, 108393–108393.

Zhao, S., Musa, S.S., Lin, Q., Ran, J., Yang, G., Wang, W., Lou, Y., Yang, L., Gao, D., He, D., Wang, M.H. (2020). Estimating the unreported number of novel coronavirus (2019-nCoV) cases in China in the first half of January 2020: A data-driven modelling analysis of the early outbreak. *Journal of Clinical Medicine*, 9(2), 388.

Zheng, Y.Y., Ma, Y.T., Zhang, J.Y. (2020). COVID-19 and the cardiovascular system. *Nature Reviews Cardiology*, 17, 259–260.

Zhou, P., Yang, XL., Wang, X.G., Hu, B., Zhang, L., Zhang, W., Si, H.R., Zhu, Y., Li, B., Huang, C.L., Chen, H.D., Chen, J., Luo, Y., Guo, H., Jiang, R.D., Liu, M.Q., Chen, Y., Shen, X.R., Wang, X., Zheng, X.S., Zhao, K., Chen, Q.J., Deng, F., Liu, L.L., Yan, B., Zhan, F.X., Wang, Y.Y., Xiao, G.F., Shi, Z.L. (2020). A pneumonia outbreak associated with a new coronavirus of probable bat origin. *Nature*, 579, 270–273.

PART 2

Applied Stochastic and Statistical Models and Methods

10

The Double Flexible Dirichlet: A Structured Mixture Model for Compositional Data

Vectors of proportions arise in a great variety of fields: chemistry, economics, medicine, sociology and many others. Supposing that a whole can be split into D mutually exclusive and exhaustive categories, vectors describing the percentage of each category on the total are referred to as compositional data. The latter are subject to a unit-sum constraint and thus their domain is the D-part simplex. A very popular distribution defined on the simplex is the Dirichlet one. This distribution, despite its several mathematical properties, is poorly parameterized and, therefore, it cannot model many dependence patterns. Some authors have proposed alternatives to the Dirichlet, looking for more flexible distributions which still retain some relevant properties for compositional data. Among these is the flexible Dirichlet (FD), introduced by Ongaro and Migliorati (2013), which generalizes the Dirichlet distribution, that is included as an inner point. Thanks to its mixture structure with D components, it exhibits a more suitable modelization of the covariance matrix. Despite its greater flexibility, the FD lacks in allowing for positive covariances, which are plausible in many applications. The aim of this contribution is to present a further generalization of the Dirichlet, called double flexible Dirichlet (DFD), that takes advantage of a finite mixture structure similar to that of the FD (depending on $D(D+1)/2$ mixture components) and enables positive covariances. Some theoretical results are shown and an estimation procedure based on the EM algorithm is proposed, including an *ad hoc* initialization strategy. A simulation study aimed at evaluating the performance of the EM algorithm under several parameter configurations is included.

Chapter written by Roberto ASCARI, Sonia MIGLIORATI and Andrea ONGARO.

10.1. Introduction

Compositional data are positive vectors subject to a unit-sum constraint, meaning that their support is the D-part simplex:

$$\mathcal{S}^D = \left\{ \mathbf{x} \in \mathbb{R}^D : x_r > 0,\ r = 1, \dots, D,\ \sum_{r=1}^{D} x_r = 1 \right\}. \qquad [10.1]$$

These data can be constructed from a basis, which is an unconstrained positive vector $\mathbf{y} = (y_1, \dots, y_D)^\intercal \in \mathbb{R}_+^D$. A composition is identified by a basis through the closure operator $\mathcal{C}(\cdot) : \mathbb{R}_+^D \to \mathcal{S}^D$, defined as:

$$\mathbf{x} = \mathcal{C}(\mathbf{y}) \equiv \mathbf{y}/y^+, \qquad [10.2]$$

where $y^+ = \sum_{r=1}^{D} y_r$ is the size of the basis $\mathbf{y}$. To model compositional data, a distribution defined on $\mathcal{S}^D$ is required. The Dirichlet distribution is the most famous simplex distribution, thanks to its several statistical properties. Nonetheless, several authors pointed out the rigid pattern it imposes on the dependence structure of the elements of composition (Aitchison 2003; Ongaro and Migliorati 2013). Let $\mathbf{X} = (X_1, \dots, X_D)^\intercal$ be a Dirichlet distributed random vector parameterized by $\boldsymbol{\alpha} = (\alpha_1, \dots, \alpha_D)^\intercal \in \mathbb{R}_+^D$ (thus, $\mathbf{X} \sim \text{Dir}(\boldsymbol{\alpha})$). The probability density function characterizing this distribution is given by:

$$f_{\text{Dir}}(\mathbf{x}; \boldsymbol{\alpha}) = \frac{\Gamma(\alpha^+)}{\prod_{j=1}^{D} \Gamma(\alpha_j)} \prod_{j=1}^{D} x_j^{\alpha_j - 1}, \qquad [10.3]$$

where $\alpha^+ = \sum_{r=1}^{D} \alpha_r$. The first two order moments of this distribution are:

$$\mathbb{E}[X_j] = \frac{\alpha_j}{\alpha^+} \qquad [10.4]$$

$$\text{Var}(X_j) = \frac{\mathbb{E}[X_j](1 - \mathbb{E}[X_j])}{\alpha^+ + 1} \qquad [10.5]$$

$$\text{Cov}(X_j, X_l) = -\frac{\mathbb{E}[X_j]\,\mathbb{E}[X_l]}{\alpha^+ + 1}, \qquad j \neq l. \qquad [10.6]$$

It is easy to see that, once the mean vector of $\mathbf{X}$ is chosen, only one parameter, namely α^+, is devoted to modeling the entire covariance matrix. In particular, if two elements of the composition have the same expected value, then they also have the same variance. Furthermore, covariances are strictly proportional to the product of the expectations of the corresponding elements. This poor parameterization does not allow for either positive covariances or multimodality, which can be an important limitation. Even though the unit-sum constraint naturally induces a negative dependence among the elements of a composition, some pairs of variables can be positively associated. For example, let us consider the composition of family-budget expenditure with categories "Food", "Clothes", "Travels" and "Other/savings". It is reasonable to think that the percentage of income spent on food can be positively associated with the percentage of income spent on clothes when the families differ in the number of members. Simpler models such as the Dirichlet fail to detect this data feature.

In the statistical literature, it is possible to find several alternatives to the Dirichlet distribution (Connor and Mosimann 1969; Gupta and Richards 1987; Barndorff-Nielsen and Jørgensen 1991; Aitchison 2003; Favaro *et al.* 2011). A recent proposal is the flexible Dirichlet (FD) (Ongaro and Migliorati 2013) that generalizes the Dirichlet distribution allowing for multimodal shapes of its probability density function and a flexible modelization of the covariance matrix. In the next section, the FD is introduced in a more rigorous way. A further completely new generalization of the Dirichlet is proposed in section 10.2, whereas in section 10.3 we show an *ad hoc* estimation procedure for this new simplex distribution.

10.1.1. *The flexible Dirichlet distribution*

The FD distribution arises closing a particular basis $\mathbf{Y} = (Y_1, \dots, Y_D)^\intercal$, i.e.:

$$Y_j = W_j + U \cdot Z_j, \qquad j = 1, \dots, D, \tag{10.7}$$

where $\mathbf{W} = (W_1, \dots, W_D)^\intercal$, U and $\mathbf{Z} = (Z_1, \dots, Z_D)^\intercal$ are jointly independent, $W_j \sim \text{gamma}(\alpha_j, 1)$, $W_j \perp\!\!\!\perp W_l$ $(j \neq l)$, $U \sim \text{gamma}(\tau, 1)$, $\mathbf{Z} \sim \text{Multinomial}(1, \mathbf{p})$, $\tau > 0$ and $\mathbf{p} \in \mathcal{S}^D$. The basis $\mathbf{Y}$ is said to have a "flexible gamma" distribution (some properties of this distribution can be found in Ongaro and Migliorati (2013)), whereas its normalized version $\mathbf{X} = \mathcal{C}(\mathbf{Y})$ is said to have a "flexible Dirichlet" distribution. A key aspect of the FD is that it can be expressed as a finite mixture of particular Dirichlet components:

$$\text{FD}(\mathbf{x}; \boldsymbol{\alpha}, \tau, \mathbf{p}) = \sum_{r=1}^{D} p_r \text{Dir}(\mathbf{x}; \boldsymbol{\alpha} + \tau \mathbf{e}_r), \tag{10.8}$$

where $\mathbf{e}_r$ is the vector with elements equal to zero except for the r-th that is equal to 1. It is well known that each mixture component can be thought of as a sub-population (cluster) in the population (Frühwirth-Schnatter 2006); therefore, the FD can include a number $k \leq D$ of different modes, one for each cluster, even if its components do not allow for multimodality. The FD also includes the Dirichlet as a special case if $\tau = 1$ and $p_r = \alpha_r / \alpha^+$, $r = 1, \dots, D$. The rich parameterization of this distribution allows for a flexible modelization of the covariance matrix of a composition, overcoming the drawbacks highlighted in section 10.1, even if covariances are still always negative.

In this chapter, we extend the FD distribution to obtain an even more flexible cluster structure and a modelization of the covariance matrix allowing for positive linear dependence.

10.2. The double flexible Dirichlet distribution

In this section, we introduce a generalization of the FD distribution, called double flexible Dirichlet (DFD). This new simplex distribution is obtained by closing a particular basis that "doubles" the generating mechanism of the FD distribution.

Let $\mathbf{W} = (W_1, \ldots, W_D)^\intercal$ be a vector of independent elements distributed as in [10.7]. Let $\mathbf{U} = (U_1, U_2)^\intercal$ be a vector of independent gamma random variables such that $U_1, U_2 \sim \text{gamma}(\tau, 1)$ and $\mathbf{Z} = (\mathbf{Z}_1, \mathbf{Z}_2)^\intercal$, where the marginals $\mathbf{Z}_1$ and $\mathbf{Z}_2$ are distributed as multinomial$(1, \mathbf{p})$ and the joint distribution is given by $p_{rh} = \mathbb{P}(\mathbf{Z}_1 = \mathbf{e}_r, \mathbf{Z}_2 = \mathbf{e}_h)$, $r, h = 1, \ldots, D$. Assuming that $\mathbf{W}, \mathbf{U}$ and $\mathbf{Z}$ are jointly independent, we can define the generic element of a new basis:

$$Y_j = W_j + U_1 \cdot Z_{1j} + U_2 \cdot Z_{2j}, \qquad j = 1, \ldots, D. \qquad [10.9]$$

The vector $\mathbf{p}$ can be thought of as the row sum (or column sum) of a symmetric matrix $\mathbf{P}$ whose generic element is p_{rh}. Then, the basis $\mathbf{Y}$ is said to have a double flexible gamma (DFG) distribution with parameters $\boldsymbol{\alpha}, \tau$ and $\mathbf{P}$. The first two order moments of this distribution are:

$$\mathbb{E}\left[Y_j\right] = \alpha_j + 2\tau p_{j\cdot} \qquad [10.10]$$

$$\text{Var}\left(Y_j\right) = \alpha_j + 2\tau p_{j\cdot} + 2\tau^2\left(p_{j\cdot} - 2p_{j\cdot}^2 + p_{jj}\right) \qquad [10.11]$$

$$\text{Cov}\left(Y_j, Y_l\right) = 2\tau^2(p_{jl} - 2p_{j\cdot}p_{l\cdot}), \qquad j \neq l \qquad [10.12]$$

where $p_{j\cdot} = \sum_{l=1}^{D} p_{jl}$. Unlike the bases characterizing both the Dirichlet and the FD distributions, the DFG allows for positively correlated elements, indeed:

$$\text{Cov}\left(Y_j, Y_l\right) \geq 0 \iff \frac{p_{j\cdot}p_{l\cdot}}{p_{jl}} \leq \frac{1}{2}. \qquad [10.13]$$

If we close a basis $\mathbf{Y} \sim \text{DFG}(\boldsymbol{\alpha}, \tau, \mathbf{P})$, then the resulting composition is said to have a DFD distribution:

$$\mathbf{X} \equiv \mathcal{C}(\mathbf{Y}) \sim \text{DFD}(\boldsymbol{\alpha}, \tau, \mathbf{P}). \qquad [10.14]$$

By conditioning on $\mathbf{Z}_1$ and $\mathbf{Z}_2$, it is possible to rewrite the DFD distribution as a finite mixture of particular Dirichlet components:

$$\text{DFD}(\mathbf{x}; \boldsymbol{\alpha}, \tau, \mathbf{P}) = \sum_{r=1}^{D}\sum_{h=1}^{D} p_{rh}\text{Dir}(\mathbf{x}; \boldsymbol{\alpha} + \tau\left(\mathbf{e}_r + \mathbf{e}_h\right)). \qquad [10.15]$$

The symmetry of $\mathbf{P}$ guarantees that only $\frac{D(D+1)}{2}$ mixture components – instead of D^2 – define the DFD distribution. Given this representation, it is easy to obtain the density function characterizing the DFD:

$$f_{DFD}(\mathbf{x};\boldsymbol{\alpha},\tau,\mathbf{P}) = \frac{\Gamma(\alpha^+ + 2\tau)}{\prod_{j=1}^{D}\Gamma(\alpha_j)}\left(\prod_{j=1}^{D} x_j^{\alpha_j - 1}\right) \cdot \left[\sum_{\substack{r=1 \\ r\neq h}}^{D}\sum_{h=1}^{D} p_{rh}\frac{\Gamma(\alpha_r)\Gamma(\alpha_h)(x_r x_h)^{\tau}}{\Gamma(\alpha_r+\tau)\Gamma(\alpha_h+\tau)} + \sum_{r=1}^{D} p_{rr}\frac{\Gamma(\alpha_r)x_r^{2\tau}}{\Gamma(\alpha_r+2\tau)}\right]$$

10.2.1. *Mixture components and cluster means*

An interesting feature of the DFD is represented by the high number of potential clusters and their position on the simplex. Indeed, if the matrix $\mathbf{P}$ is symmetric with each $p_{rh} > 0$, then $D^* = \frac{D(D+1)}{2}$ different clusters are potentially present. Since the generic component is distributed as a $\text{Dir}(\boldsymbol{\alpha} + \tau\,(\mathbf{e}_r + \mathbf{e}_h))$, $r, h = 1, \ldots, D$, its mean vector is:

$$\begin{aligned}\boldsymbol{\mu}_{rh} &= \frac{\boldsymbol{\alpha} + \tau\,(\mathbf{e}_r + \mathbf{e}_h)}{\alpha^+ + 2\tau} \\ &= \left(\frac{\alpha^+}{\alpha^+ + 2\,\tau}\right)\bar{\boldsymbol{\alpha}} + \left(\frac{\tau}{\alpha^+ + 2\,\tau}\right)\mathbf{e}_r + \left(\frac{\tau}{\alpha^+ + 2\,\tau}\right)\mathbf{e}_h,\end{aligned} \qquad [10.16]$$

where $r, h = 1, \ldots, D$, $r \leq h$ and $\bar{\boldsymbol{\alpha}} = \boldsymbol{\alpha}/\alpha^+$. Each $\boldsymbol{\mu}_{rh}$ can be expressed as the weighted mean of three quantities: a common barycenter $\bar{\boldsymbol{\alpha}}$ and the two vertices of the simplex $\mathbf{e}_r$ and $\mathbf{e}_h$. These cluster means (and, consequently, the corresponding mixture components) are located in a precise scheme on the simplex, as shown in Figure 10.1, where blue triangles represent the cluster means and the green one represents $\bar{\boldsymbol{\alpha}}$ in a scenario with $D = 3$. If we connect the cluster means, we obtain a "scaled simplex" with edges parallel to the simplex ones. Vertices of this scaled simplex are $\boldsymbol{\mu}_{11}$, $\boldsymbol{\mu}_{22}$ and $\boldsymbol{\mu}_{33}$. Vectors $\boldsymbol{\mu}_{rh}$ with $r \neq h$ are situated at the midpoint of the segment joining $\boldsymbol{\mu}_{rr}$ and $\boldsymbol{\mu}_{hh}$.

From equation [10.16] and Figure 10.1, it is easy to see that the parameter τ regulates the distance between each cluster barycenter and $\bar{\boldsymbol{\alpha}}$: increasing τ, we obtain cluster barycenters closer to the simplex boundary. While the above structure is somewhat rigid, it is similar to that of the FD distribution (Migliorati *et al*. 2017) allowing for more clusters. Furthermore, thanks to the fact that some p_{rh} can be equal to 0, this model allows for a variety of cluster configurations that cannot be obtained with the FD model. For example, in Figure 10.2, it is possible to see some cluster configurations that cannot be reached by simpler models. Please note that joining the cluster means in these two panels produces a diamond and an inverse triangle shape, respectively.

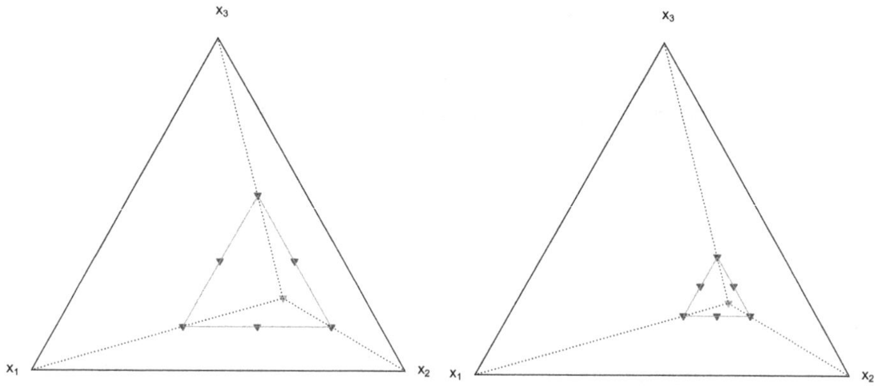

Figure 10.1. *DFD cluster means structure.* $\alpha = (5, 13, 5)^{\top}$*. Left:* $\tau = 15$*, right:* $\tau = 5$*. For a color version of this figure, see www.iste.co.uk/dimotikalis/analysis2.zip*

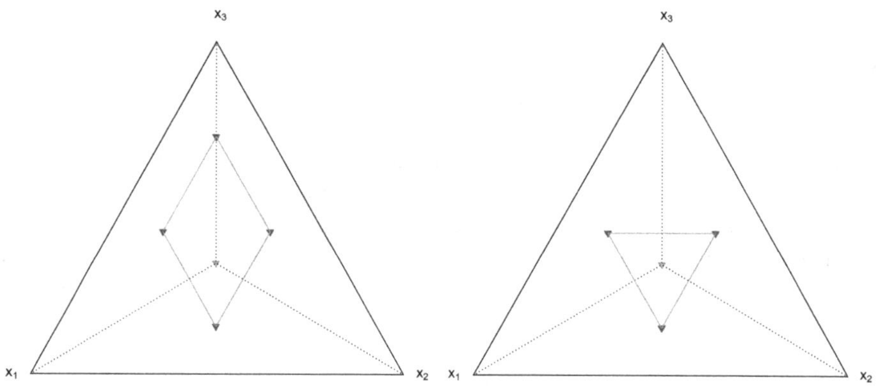

Figure 10.2. *DFD cluster means with* $\alpha = (5, 5, 5)^{\top}$ *and* $\tau = 10$*. Left:* $p_{11} = p_{22} = 0$*. Right:* $p_{11} = p_{22} = p_{33} = 0$*. For a color version of this figure, see www.iste.co.uk/dimotikalis/analysis2.zip*

Thanks to the mixture representation, computing the first two order moments of the DFD distribution is easy:

$$\mathbb{E}\left[X_j\right] = \frac{\alpha_j + 2\tau\, p_{j\cdot}}{\alpha^+ + 2\tau} \qquad [10.17]$$

$$\text{Var}\left(X_j\right) = \frac{\mathbb{E}\left[X_j\right]\left(1 - \mathbb{E}\left[X_j\right]\right)}{\alpha^+ + 2\tau + 1} + \frac{2\tau^2\left[p_{j\cdot}(1 - 2p_{j\cdot}) + p_{jj}\right]}{(\alpha^+ + 2\tau + 1)(\alpha^+ + 2\tau)} \qquad [10.18]$$

$$\text{Cov}\left(X_j, X_l\right) = -\frac{\mathbb{E}\left[X_j\right]\mathbb{E}\left[X_l\right]}{\alpha^+ + 2\tau + 1} + \frac{2\tau^2(p_{jl} - 2p_{j\cdot}p_{l\cdot})}{(\alpha^+ + 2\tau + 1)(\alpha^+ + 2\tau)} \quad j \neq l. \qquad [10.19]$$

In general, the covariance matrix of a basis is rarely connected to the covariance matrix of a composition in a simple way. The most simple but effective example to show this aspect is the closure of a basis formed by independent gamma random variables leading to always negative covariances (the Dirichlet ones). In other words, no automatic relation between basis and composition dependence structure exists; it instead depends on the underlying basis distribution. An interesting feature of the DFD model is that the dependence induced in the basis by the generating mechanism [10.9] appears in the composition. Indeed, the covariance among two elements of a DFD-distributed vector can be written as:

$$\text{Cov}\left(X_j, X_l\right) = -\frac{\mathbb{E}\left[X_j\right]\mathbb{E}\left[X_l\right]}{\alpha^+ + 2\tau + 1} + \frac{\text{Cov}\left(Y_j, Y_l\right)}{(\alpha^+ + 2\tau + 1)(\alpha^+ + 2\tau)}, \quad j \neq l. \qquad [10.20]$$

The first element is always negative and it is due to the closure of a gamma-related basis, whereas the second term is exactly the covariance of the corresponding basis elements divided by a constant. Since this last term can assume both positive and negative values according to the difference $(p_{jl} - 2p_{j\cdot}p_{l\cdot})$, it influences the negative linear dependence which is typical of the Dirichlet. In particular, thanks to this new term, the covariance among two components can assume values greater than zero, allowing for positive dependence.

Even if the analytical expression for Pearson's correlation coefficient ρ_{jl} of two arbitrary components X_j and X_l $(j, l = 1, \ldots, D, j \neq l)$ of **X** is hardly tractable, it is easy to show that it may take high positive values.

EXAMPLE 10.1.– *Let us consider the matrix* **P** *satisfying the following constraints:*

$$\begin{cases} p_{jl} = p_{lj} = \frac{1}{4} \\ p_{j\cdot} = p_{l\cdot} = \frac{1}{4} \\ p_{jj} = p_{ll} = 0 \end{cases} . \qquad [10.21]$$

Then, the following quantities can be computed:

$$\mathbb{E}[X_q] = \frac{\alpha_q + \frac{\tau}{2}}{\alpha^+ + 2\tau} \underset{\tau \to +\infty}{\longrightarrow} \frac{1}{4}, \quad q \in \{j, l\}$$

$$Var(X_q) = \frac{\mathbb{E}[X_q](1 - \mathbb{E}[X_q])}{\alpha^+ + 2\tau + 1} + \frac{2\tau^2 \frac{1}{8}}{(\alpha^+ + 2\tau + 1)(\alpha^+ + 2\tau)} \underset{\tau \to +\infty}{\longrightarrow} \frac{1}{16}, \quad q \in \{j, l\}$$

$$Cov(X_j, X_l) = -\frac{\mathbb{E}[X_j]\mathbb{E}[X_l]}{\alpha^+ + 2\tau + 1} + \frac{2\tau^2 \frac{1}{8}}{(\alpha^+ + 2\tau + 1)(\alpha^+ + 2\tau)} \underset{\tau \to +\infty}{\longrightarrow} -0 + \frac{1}{16} = \frac{1}{16}$$

$$\Longrightarrow \rho_{jl} = \frac{Cov(X_j, X_l)}{\sqrt{Var(X_j) \cdot Var(X_l)}} \underset{\tau \to +\infty}{\longrightarrow} \frac{\frac{1}{16}}{\sqrt{\frac{1}{16} \cdot \frac{1}{16}}} = 1.$$

10.3. Computational and estimation issues

In the previous section, the DFD distribution has been introduced and some theoretical properties have been listed. In this section, the interest is in providing an estimation procedure for the parameters $\boldsymbol{\alpha}$, τ and $\mathbf{P}$. To this end, it is useful to define a cluster-code matrix:

DEFINITION 10.1.– *A* ***cluster-code matrix*** *of order* D, $\boldsymbol{C}_D \in \mathcal{M}(D, D)$, *is an upper triangular matrix such that:*

– the main diagonal is composed of the first (ordered) D integers:

$$\boldsymbol{C}_5 = \begin{bmatrix} 1 & \cdot & \cdot & \cdot & \cdot \\ & 2 & \cdot & \cdot & \cdot \\ & & 3 & \cdot & \cdot \\ & & & 4 & \cdot \\ & & & & 5 \end{bmatrix}$$

– the remaining elements are equal to the (ordered) integers from $D + 1$ *to* $D^* = \dfrac{D(D+1)}{2}$ *allocated by row:*

$$\boldsymbol{C}_5 = \begin{bmatrix} 1 & 6 & 7 & 8 & 9 \\ & 2 & \cdot & \cdot & \cdot \\ & & 3 & \cdot & \cdot \\ & & & 4 & \cdot \\ & & & & 5 \end{bmatrix} \Rightarrow \begin{bmatrix} 1 & 6 & 7 & 8 & 9 \\ & 2 & 10 & 11 & 12 \\ & & 3 & \cdot & \cdot \\ & & & 4 & \cdot \\ & & & & 5 \end{bmatrix} \Rightarrow \begin{bmatrix} 1 & 6 & 7 & 8 & 9 \\ & 2 & 10 & 11 & 12 \\ & & 3 & 13 & 14 \\ & & & 4 & 15 \\ & & & & 5 \end{bmatrix}$$

Given a particular value of D, a cluster-code matrix allows us to identify a particular cluster uniquely. Let $k \in \{1, 2, \ldots, D^*\}$ be a cluster label; if $c_{rh} = k$, then cluster k is the one distributed as a Dirichlet with parameters $\boldsymbol{\alpha} + \tau(\mathbf{e}_r + \mathbf{e}_h)$. With this cluster structure, the DFD model can be rewritten as:

$$\text{DFD}(\mathbf{x}; \boldsymbol{\alpha}, \tau, \boldsymbol{\pi}) = \sum_{k=1}^{D^*} \pi_k \text{Dir}(\mathbf{x}; \boldsymbol{\alpha} + \tau e(k)),$$

where $e(k) = \sum_{r=1}^{D}\sum_{h\geq r}(\mathbf{e}_r + \mathbf{e}_h)\cdot I(c_{rh} = k)$, $\boldsymbol{\pi} = (\pi_1, \ldots, \pi_{D^*})^\intercal$ and

$$\pi_k = \begin{cases} p_{kk} & \text{if } k = 1, \ldots, D \\ 2p_{\{rh\,:\,c_{rh}=k\}} & \text{if } k = D+1, \ldots, D^* \end{cases}.$$

This new notation makes the definition of the cluster barycenters easier:

$$\boldsymbol{\mu}_k = \left(\frac{\alpha^+}{\alpha^+ + 2\,\tau}\right)\bar{\boldsymbol{\alpha}} + \left(\frac{\tau}{\alpha^+ + 2\,\tau}\right)e(k), \qquad k = 1, \ldots, D^*. \qquad [10.22]$$

10.3.1. *Parameter estimation: the EM algorithm*

Let us assume that a random sample $\mathbf{x} = (\mathbf{x}_1, \mathbf{x}_2, \ldots, \mathbf{x}_n)^\intercal$ of size n has been collected, where each $\mathbf{x}_i$ $(i = 1, \ldots, n)$ is a realization from $\mathbf{X} \sim \text{DFD}(\boldsymbol{\alpha}, \tau, \boldsymbol{\pi})$. To obtain the value of $\boldsymbol{\theta} = (\boldsymbol{\alpha}, \tau, \boldsymbol{\pi})^\intercal$ that maximizes the likelihood function, we may use the expectation-maximization (EM) algorithm, formalized by Dempster *et al.* (1977). In this context, the EM algorithm is defined to maximize the conditional expectation of the complete-data log likelihood function:

$$\log L_C(\boldsymbol{\theta}|\mathbf{x}) = \sum_{i=1}^{n}\sum_{k=1}^{D^*} z_{ik}\left\{\log \pi_k + \log f_D(\mathbf{x}_i; \boldsymbol{\alpha} + \tau \cdot e(k))\right\}, \qquad [10.23]$$

where z_{ik} is a component indicator that is equal to 1 if observation $\mathbf{x}_i$ has arisen from cluster k.

It is well known that the EM algorithm is not robust with respect to the choice of the initial values (Diebolt and Ip 1996; Biernacki *et al.* 2003; O'Hagan *et al.* 2012). For this reason, an *ad hoc* initialization procedure has been implemented. It requires a partition of the sample $\mathbf{x} = (\mathbf{x}_1, \ldots, \mathbf{x}_n)^\intercal$ into D^* groups and thus a clustering method. A hierarchical clustering based on the Aitchison metric (Pawlowsky-Glahn and Egozcue 2002) and four k-means algorithms based on different transformations of the compositions have been compared. An exploratory simulation study has highlighted that the k-means algorithm based on the entire untransformed compositions works better in most parameter configurations. Although in the DFD context there exists a clear cluster structure, the k-means algorithm (as any clustering method) labels clusters in a random way. Thus, a labeling scheme has been *ad hoc* constructed to assign the "correct" label to each cluster. Suppose, without loss of generality, that $D = 3$ so that $D^* = 6$. Remembering that the component-specific distribution is $\text{Dir}(\boldsymbol{\alpha} + \tau e(k))$, the mean vector for each cluster can be expressed as in Table 10.1.

Cluster k	Component μ_{k1}	μ_{k2}	μ_{k3}
1	$\frac{\alpha_1+2\tau}{\alpha^+ + 2\tau}$	$\frac{\alpha_2}{\alpha^+ + 2\tau}$	$\frac{\alpha_3}{\alpha^+ + 2\tau}$
2	$\frac{\alpha_1}{\alpha^+ + 2\tau}$	$\frac{\alpha_2+2\tau}{\alpha^+ + 2\tau}$	$\frac{\alpha_3}{\alpha^+ + 2\tau}$
3	$\frac{\alpha_1}{\alpha^+ + 2\tau}$	$\frac{\alpha_2}{\alpha^+ + 2\tau}$	$\frac{\alpha_3+2\tau}{\alpha^+ + 2\tau}$
4	$\frac{\alpha_1+\tau}{\alpha^+ + 2\tau}$	$\frac{\alpha_2+\tau}{\alpha^+ + 2\tau}$	$\frac{\alpha_3}{\alpha^+ + 2\tau}$
5	$\frac{\alpha_1+\tau}{\alpha^+ + 2\tau}$	$\frac{\alpha_2}{\alpha^+ + 2\tau}$	$\frac{\alpha_3+\tau}{\alpha^+ + 2\tau}$
6	$\frac{\alpha_1}{\alpha^+ + 2\tau}$	$\frac{\alpha_2+\tau}{\alpha^+ + 2\tau}$	$\frac{\alpha_3+\tau}{\alpha^+ + 2\tau}$

Table 10.1. *Mean vectors stratified by cluster. μ_{kj} refers to the j-th element of $\boldsymbol{\mu}_k$. For a color version of this figure, see www.iste.co.uk/dimotikalis/analysis2.zip*

It is easy to note that the highest value of μ_{kj} (the j-th element of $\boldsymbol{\mu}_k$) is reached when $k = j$, as we can see from the red fractions in Table 10.1. If we estimate each μ_{kj} with the cluster sample mean $\bar{x}_{kj}$, we can label the cluster associated with the greatest $\bar{x}_{kj}$ ($j = 1, \dots, D$) as cluster j. To label the remaining clusters, let us consider the set of indices $\mathcal{U}_r$:

$$\mathcal{U}_r = \{k : c_{hr} = k,\ h = 1, \dots, r \qquad \& \qquad k : c_{rh} = k,\ h = r+1, \dots, D\}. \tag{10.24}$$

If $k \in \mathcal{U}_r$, then index k is on the r-th row or the r-th column of the cluster code matrix $\mathbf{C}_D$. Conditioning on cluster $k > D$, it is easy to note that μ_{kr} is maximized by those $k \in \mathcal{U}_r \setminus \{r\}$ (fractions in blue).

Then, the cluster that maximizes $\mu_{\cdot r}$ and $\mu_{\cdot h}$ is labeled as $k = c_{rh}$. If multiple label schemes occur, the estimation procedure is applied to every single label permutation compatible with the observed structure. Given a data partition obtained with the above method, an initialization for $\boldsymbol{\pi}$ is the percentage of data points allocated to each cluster:

$$\boldsymbol{\pi}^{(0)} = \left(\pi_1^{(0)}, \dots, \pi_{D^*}^{(0)}\right)^\mathsf{T},$$

where $\pi_k^{(0)} = \frac{1}{n}\sum_{i=1}^n \hat{z}_{ik}$ and $\hat{z}_{ik}$ is the sample version of z_{ik}, i.e. it is an indicator that is equal to 1 if observation $\mathbf{x}_i$ has been allocated to cluster k.

To obtain initializations of $\boldsymbol{\alpha}$ and τ, the method of moments based on all the D^* clusters can be used. Let $\bar{x}_{rj}$ and s_{rj}^2 be the sample mean and the sample variance of component j among cluster r. Remembering that $\mathbb{E}\left[X_{rj}\right] = \frac{\alpha_h}{\alpha^+ + 2\tau}$ if $r \not\in \mathcal{U}_j$, then the algorithm can be expressed as:

1) Initialize $\widehat{\left(\frac{\alpha_j}{\alpha^+ + 2\tau}\right)} = \dfrac{\sum_{\substack{r=1 \\ r \notin \mathcal{U}_j}}^{D^*} \bar{x}_{rj} \cdot \pi_r^{(0)}}{\sum_{\substack{r=1 \\ r \notin \mathcal{U}_j}}^{D^*} \pi_r^{(0)}}.$

2) Given that

$$\left(\frac{2\tau}{\alpha^+ + 2\tau}\right) =$$

$$\begin{cases} \left(\dfrac{\alpha_r + 2\tau}{\alpha^+ + 2\tau}\right) - \left(\dfrac{\alpha_r}{\alpha^+ + 2\tau}\right), & \text{if } r = 1, \ldots, D \\ \left(\dfrac{\alpha_l + \tau}{\alpha^+ + 2\tau}\right) - \left(\dfrac{\alpha_l}{\alpha^+ + 2\tau}\right) + \left(\dfrac{\alpha_w + \tau}{\alpha^+ + 2\tau}\right) - \left(\dfrac{\alpha_w}{\alpha^+ + 2\tau}\right), & \text{if } r = D+1, \ldots, D^* \end{cases}$$

where l and w are two indices such that $c_{lw} = r$ or $c_{wl} = r$, then the initialization $\widehat{\left(\frac{2\tau}{\alpha^+ + 2\tau}\right)}$ is the weighted mean of the D^* quantities:

$$\begin{cases} \bar{x}_{rr} - \widehat{\left(\dfrac{\alpha_r}{\alpha^+ + 2\tau}\right)}, & \text{if } r = 1, \ldots, D \\ \bar{x}_{rl} - \widehat{\left(\dfrac{\alpha_l}{\alpha^+ + 2\tau}\right)} + \bar{x}_{rw} - \widehat{\left(\dfrac{\alpha_w}{\alpha^+ + 2\tau}\right)}, & \text{if } r = D+1, \ldots, D^* \end{cases}$$

3) Since we have obtained initializations for the "relative" counterparts of $\boldsymbol{\alpha}$ and τ, we need an initialization for the common denominator $(\alpha^+ + 2\tau)$. Remembering that $\text{Var}\left(X_{rj}\right) = \dfrac{\mathbb{E}\left[X_{rj}\right]\left(1 - \mathbb{E}\left[X_{rj}\right]\right)}{\alpha^+ + 2\tau + 1}$, initialization of $(\alpha^+ + 2\tau)$ can be obtained as the weighted mean of:

$$\frac{1 - \sum_{j=1}^{D} \bar{x}_{rj}^2}{\sum_{j=1}^{D} s_{rj}^2} - 1, \qquad r = 1, \ldots, D^*.$$

We can use $\boldsymbol{\pi}^{(0)}$ as weights in steps 2 and 3. Table 10.2 reports the means of 500 initializations for $\boldsymbol{\alpha}$ and τ. These initializations have been obtained with samples of size 300 generated from a subset of the parameter configurations reported in Table 10.3.

	α_1	α_2	α_3	τ
True	10	10	10	10
Init.	9.427	9.286	9.526	9.703
True	10	10	10	40
Init.	9.706	9.397	9.874	37.745
True	2	23	12	17
Init.	1.888	20.615	10.861	15.836

	α_1	α_2	α_3	τ
True	100	40	40	15
Init.	92.450	37.019	37.261	13.209
True	10	100	14	8
Init.	10.034	98.212	13.570	7.800
True	12	0.900	30	20
Init.	12.478	0.892	32.036	21.476

Table 10.2. *Mean of 500 initializations of (α, τ) in different parameter configurations*

ID	α_1	α_2	α_3	τ	π_1	π_2	π_3	π_4	π_5	π_6
1	10	10	10	15	0.11	0.11	0.11	0.22	0.22	0.22
2	10	10	10	40	0.11	0.11	0.11	0.22	0.22	0.22
3	2	23	12	17	0.08	0.16	0.18	0.10	0.40	0.08
4	40	20	30	25	0.00	0.16	0.26	0.40	0.00	0.18
5	40	20	30	50	0.00	0.16	0.26	0.40	0.00	0.18
6	100	40	40	15	0.22	0.17	0.15	0.15	0.10	0.20
7	40	20	30	18	0.00	0.00	0.00	0.30	0.19	0.51
8	10	100	14	8	0.10	0.15	0.15	0.10	0.40	0.10
9	12	0.90	30	20	0.08	0.16	0.18	0.10	0.40	0.08

Table 10.3. *Parameter configurations for all the DFD simulations*

10.3.2. *Simulation study*

In this section, we propose two simulation studies aimed at evaluating the performances of the estimation procedure, both of them based on the nine parametric configurations of the DFD distribution reported in Table 10.3. These configurations allow us to cover several scenarios: well separated (configurations 1–5, 7 and 9) as well as overlapping clusters (6 and 8), clusters that are very close to one edge of the simplex (3 and 9), positive (4 and 5) and negative correlations, as well as configurations with some empty components (4, 5 and 7).

The EM algorithm is one of the most popular algorithms used to maximize the likelihood function in a missing data scenario. Due to its sensitivity to initial values, several authors have proposed alternative versions of this algorithm (Celeux and Govaert 1992; Celeux *et al.* 1995; Biernacki *et al.* 2003): two of them are the classification EM (CEM) and the stochastic EM (SEM). The CEM has a further classification step: at step m, each observation is allocated to the group k maximizing $\hat{\pi}_k^{(m)}(\mathbf{x}_i; \boldsymbol{\theta})$, defined as in equation [10.25]

$$\hat{\pi}_k(\mathbf{x}_i; \boldsymbol{\theta}) = \frac{\hat{\pi}_k \cdot f_D(\mathbf{x}_i; \boldsymbol{\alpha} + \tau \cdot e(k))}{\sum_{h=1}^{D^*} \hat{\pi}_h \cdot f_D(\mathbf{x}_i; \boldsymbol{\alpha} + \tau \cdot e(h))}, \qquad k = 1, \ldots, D^*. \qquad [10.25]$$

The SEM algorithm has a similar approach: at each step, observation $\mathbf{x}_i$ is allocated to a new group according to a multinomial distribution with probabilities $\hat{\boldsymbol{\pi}}_i = (\hat{\pi}_1(\mathbf{x}_i;\boldsymbol{\theta}), \ldots, \hat{\pi}_{D^*}(\mathbf{x}_i;\boldsymbol{\theta}))^\intercal$, with components defined as in [10.25].

Given the initialization procedure we developed before, we set up a simulation study aimed at comparing different EM-based algorithms. It is composed of the following steps:

– For each configuration of parameters in Table 10.3, 100 samples of size 100 have been generated.

– For each sample, initialization of the parameters has been obtained according to the method described in section 10.3.1.

– The likelihood function has been maximized with the EM, CEM, SEM, CEM+EM and SEM+EM algorithms.

CEM+EM and SEM+EM consist of initializing the CEM/SEM with the proposed procedure and then use their results as initial values for the standard EM. In this way, we give the EM algorithm a chance to move away from a path of convergence to a local maximizer. Table 10.4 reports the proportion of simulations (column "%") where each method provided the highest log-likelihood and the mean of the log-likelihoods evaluated at the obtained final estimates. From these results, one can conclude that the SEM + EM combination is the one providing the best values in most cases. The presence of an EM step is fundamental: the CEM and the SEM are not able to find the global maximizer by themselves (look at columns "%" for the CEM and SEM methods).

	EM		CEM		SEM		CEM+EM		SEM+EM	
ID	%	Mean $\hat{l}$	%	Mean $\hat{l}$	%	Mean $\hat{l}$	%	Mean $\hat{l}$	%	Mean $\hat{l}$
1	0.287	128.6434	0	104.5987	0	104.5987	0.330	128.6434	0.383	129.5926
2	0.005	191.0407	0	189.1948	0	189.1948	0.000	191.0407	0.995	191.0408
3	0.285	169.1768	0	166.2929	0	166.2929	0.278	169.1768	0.437	169.1559
4	0.280	220.7419	0	211.3169	0	211.3169	0.330	220.7419	0.390	220.7137
5	0.018	261.2633	0	251.4802	0	251.4802	0.028	261.2633	0.953	261.2633
6	0.358	307.6959	0	279.7145	0	279.7145	0.325	307.6959	0.317	307.6959
7	0.337	216.5530	0	184.9596	0	184.9596	0.447	218.6098	0.217	216.2881
8	0.318	344.0424	0	305.5102	0	305.5102	0.330	344.0424	0.352	344.0383
9	0.295	260.0346	0	258.3536	0	258.3536	0.270	260.0346	0.435	260.0347

Table 10.4. *Results for the simulation study regarding the initialization procedure*

The second simulation study regards the evaluation of the performance of the final estimation procedure, composed of the initialization followed by the SEM+EM algorithm. For each configuration reported in Table 10.3, 1000 samples of size $n = 150$ have been generated. For each of them, the parameters of the DFD model have been estimated according to the initialization and estimation procedures described in section 10.3.1. Table 10.5 reports the results of the simulation for two particular scenarios (third and fourth ID, Figure 10.3). This table contains the true value of the parameters, the mean of the 1000 estimates for each parameter and the absolute relative bias (ARB, defined as the mean of the absolute differences between the true parameter and its estimate, divided by the true value of the parameter). Finally, we reported two quantities: the first one is the standard deviation of the 1000 estimates for each parameter, which can be thought of as the bootstrap approximation of the standard error (SE) of the estimator and, therefore, it has been called "Boot. SE". The last quantity is the coverage of the approximated 95% confidence intervals (CI, computed as $\hat{\theta} \pm z_{.975} \cdot \text{SE}_{Boot}$), which is the percentage of times that the approximated 95% CI contains the true value of the parameter. In general, it is reported that estimating parameters of a finite mixture model through the EM algorithm can encounter several issues, particularly when the sample size is small (McLachlan and Peel 2004; Frühwirth-Schnatter 2006). In the considered simulations, the relatively small sample size (fixed and equal to $n = 150$) seems to be large enough to produce very good results. In most of the scenarios, the coverage level of approximated confidence intervals is very close to the 95% nominal one.

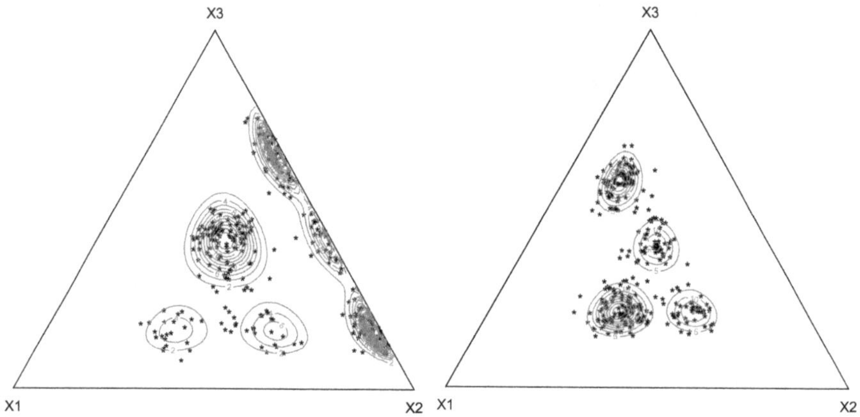

Figure 10.3. *Ternary diagrams with isodensity contour plot of the true density function of scenarios ID = 3 (left) and ID = 4 (right). For a color version of this figure, see www.iste.co.uk/dimotikalis/analysis2.zip*

As remarked in section 10.2.1, setting some null mixing weights could lead to a very interesting and particular configuration of clusters' barycenters. For example, scenario 4 is characterized by two weights equal to zero ($\pi_1 = \pi_5 = 0$) and joining the clusters' means produces an oblique and rotated "L". Scenario 3 has clusters very close to one of the edges of the simplex; this means that many observations have at least one component close to 0. This can be a problem in compositional data analysis since the presence of zeros is a well-documented issue (Aitchison 2003). Results of our simulation study confirm that if clusters are distinguishable, the estimation procedure is very reliable. Nonetheless, when clusters are overlapped, the EM-based algorithm is not able to produce good estimates. For example, scenarios 6 and 8 are quite challenging since the EM algorithm is not capable of recognizing the true number of clusters due to their overlapping. For these scenarios, the results suggest that our estimation procedure is not very precise. The same holds for many other estimation procedures applied to finite mixture characterized by indistinguishable clusters (Frühwirth-Schnatter 2006).

ID 3	α_1	α_2	α_3	τ	π_1	π_2	π_3	π_4	π_5	π_6
True	2	23	12	17	0.080	0.160	0.180	0.100	0.400	0.080
MLE mean	2.019	23.269	12.151	17.083	0.083	0.163	0.184	0.098	0.397	0.076
Arb	0.010	0.012	0.013	0.005	0.032	0.016	0.020	0.020	0.008	0.044
Boot. SE	0.206	2.022	1.091	1.567	0.024	0.030	0.033	0.025	0.040	0.022
Coverage	0.946	0.944	0.947	0.946	0.962	0.948	0.941	0.945	0.945	0.948
ID 4	α_1	α_2	α_3	τ	π_1	π_2	π_3	π_4	π_5	π_6
True	40	20	30	25	0	0.160	0.260	0.400	0	0.180
MLE mean	40.802	20.380	30.587	25.385	0	0.162	0.261	0.397	0	0.180
Arb	0.020	0.019	0.020	0.015	-	0.015	0.004	0.007	-	0.003
Boot. SE	3.245	1.717	2.477	2.114	0	0.031	0.036	0.041	0.001	0.032
Coverage	0.935	0.930	0.935	0.943	0.996	0.956	0.947	0.953	0.994	0.954

Table 10.5. *ID 4 – Simulation results*

Since the simulation study presented here was aimed only at evaluating the performances of the estimation procedure, future works will compare the DFD distribution with other popular simplex distributions in terms of fit to real and simulated data.

10.4. References

Aitchison, J. (2003). *The Statistical Analysis of Compositional Data*. The Blackburn Press, London.

Barndorff-Nielsen, O.E. and Jørgensen, B. (1991). Some parametric models on the simplex. *Journal of Multivariate Analysis*, 39(1), 106–116.

Biernacki, C., Celeux, G., Govaert, G. (2003). Choosing starting values for the EM algorithm for getting the highest likelihood in multivariate Gaussian mixture models. *Computational Statistics & Data Analysis*, 41, 561–575.

Celeux, G. and Govaert, G. (1992). A classification EM algorithm for clustering and two stochastic versions. *Computational Statistics & Data Analysis – Special Issue on Optimization Techniques in Statistics*, 14(3), 315–332.

Celeux, G., Chauveau, D., Diebolt, J. (1995). On stochastic versions of the EM algorithm on stochastic versions of the EM algorithm. Technical report, INRIA.

Connor, R. and Mosimann, J.E. (1969). Concepts of independence for proportions with a generalization of the Dirichlet distribution. *Journal of the American Statistical Association*, 64(325), 194–206.

Dempster, A.P., Laird, N.M., Rubin, D.B. (1977). Maximum likelihood from incomplete data via the EM algorithm. *Journal of the Royal Statistical Society, Series B: Methodological*, 39(1), 1–38.

Diebolt, J. and Ip, E.H.S. (1996). Stochastic EM: Method and application. *Markov Chain Monte Carlo in Practice*, 259–273.

Favaro, S., Hadjicharalambous, G., Prünster, I. (2011). On a class of distributions on the simplex. *Journal of Statistical Planning and Inference*, 141(9), 2987–3004.

Frühwirth-Schnatter, S. (2006). *Finite Mixture and Markov Switching Models*. Springer, New York.

Gupta, R.D. and Richards, D.St.P. (1987). Multivariate Liouville distributions. *Journal of Multivariate Analysis*, 23, 233–256.

McLachlan, G. and Peel, D. (2004). *Finite Mixture Models*. John Wiley & Sons, New York.

Migliorati, S., Ongaro, A., Monti, G.S. (2017). A structured Dirichlet mixture model for compositional data: Inferential and applicative issues. *Statistics and Computing*, 27(4), 963–983.

O'Hagan, A., Murphy, T.B., Gormley, I.C. (2012). Computational aspects of fitting mixture models via the expectation–maximization algorithm. *Computational Statistics and Data Analysis*, 56(12), 3843–3864.

Ongaro, A. and Migliorati, S. (2013). A generalization of the dirichlet distribution. *Journal of Multivariate Analysis*, 114(1), 412–426.

Pawlowsky-Glahn, V. and Egozcue, J.J. (2002). BLU estimators and compositional data. *Mathematical Geology*, 34(3), 259–274.

11

Quantization of Transformed Lévy Measures

In this chapter, we find an optimal approximation of the measure associated with a transformed version of the Lévy–Khintchine canonical representation via a convex combination of a finite number P of Dirac masses. The quality of such an approximation is measured in terms of the Monge–Kantorovich, known also as the Wasserstein metric. In essence, this procedure is equivalent to the quantization of measures. This method requires prior knowledge of the functional form of the measure. However, since this is in general not known, then we shall have to estimate it. It will be shown that the objective function used to estimate the position of the Dirac masses and their associated weights (or masses) can be expressed as a stochastic program. The properties of the estimator provided are discussed. Also, a number of simulations for different types of Lévy processes are performed and the results are discussed.

11.1. Introduction

Recently, there has been a sharp rise of interest in the study of Lévy processes. This is because their applications are far-reaching. These processes have been applied in various fields of research which include telecommunications, quantum theory, extreme value theory, insurance and finance. Lévy processes can be defined as stochastic processes that are stochastically continuous, with increments that are independent and stationary. Moreover, it is possible to find a version of such a process that is almost surely right continuous with left limits.

The relation between Lévy processes and the family of infinitely divisible distributions has been studied extensively and has been well documented. In fact,

Chapter written by Mark Anthony CARUANA.

there is a one-to-one relation between infinitely divisible distributions and Lévy processes. A number of texts which include amongst others (Bertoin 1996; Sato 1999; Applebaum 2004; Kyprianou 2006) illustrate and discuss the fundamental properties of Lévy processes.

The characteristic function of an infinitely divisible distribution which we denote by $\varphi(t)$, and which is associated with a specific Lévy process, has a number of different representations which include the celebrated Lévy–Khintchine representation, the Lévy–Khintchine canonical representation, the Lévy canonical representation and the Kolmogorov representation. Recently, in Sant and Caruana (2017), another representation was developed. Through this representation, the characteristic function of the increments of a Lévy process can be expressed as follows:

$$\varphi(t) = \exp\left(i\gamma t + \int_{\mathbb{R}} \left(\exp(itu) - 1 - \frac{itu}{1+|u|^{2-|u|^{\beta}}}\right)\left(\frac{1+|u|^{2-|u|^{\beta}}}{|u|^{2-|u|^{\beta}}}\right) dG(u)\right), \quad [11.1]$$

where γ is the drift term and G is a non-decreasing function of bounded variation such that $G(-\infty) = 0$ and $G(\infty) < \infty$. Furthermore, we have that

$$\lim_{u\to 0}\left[\left(\exp(itu) - 1 - \frac{itu}{1+|u|^{2-|u|^{\beta}}}\right)\left(\frac{1+|u|^{2-|u|^{\beta}}}{|u|^{2-|u|^{\beta}}}\right)\right] = \frac{-t^2}{2}. \quad [11.2]$$

The parameter γ and the function G together completely determine the infinitely divisible distribution, and thus, they allow us to identify a Lévy process. This representation is similar to the so-called Lévy–Khintchine canonical representation in which $\varphi(t)$ can be expressed as follows:

$$\varphi(t) = \exp\left(i\delta t + \int_{\mathbb{R}} \left(\exp(itu) - 1 - \frac{itu}{1+u^2}\right)\left(\frac{1+u^2}{u^2}\right) dH(u)\right). \quad [11.3]$$

Like G, the function H is also non-decreasing and with bounded variation such that $H(-\infty) = 0$ and $H(\infty) < \infty$. Sant and Caruana (2017) also discuss the relation between the functions H, G and the Lévy measure which features in the Lévy–Khintchine representation and is usually denoted by $v(.)$. In this chapter, we will primarily focus our efforts on the estimation of the measure associated with the function G and which is defined in [11.1]. In particular, we will assume that G is continuous except at the origin. At this point, the function G and also the function H both experience a jump. This jump is caused by the Brownian motion component. The literature related to the parameter estimation of Lévy processes is primarily divided into two approaches: the parametric and the non-parametric.

However, the semi-parametric approach is also considered. The parametric approach is chosen if we decide to fully parametrize the function G. Typical examples of Lévy processes which fit within this framework include and are not limited to the gamma process, the Cauchy process, the Carr–Geman–Madan–Yor (CGMY) process and the Stable process. Over the years, a number of techniques have been devised to estimate the parameters of these processes, these include, among others, the method of maximum likelihood estimation (MLE), the generalized method of moments (GMM), the method of maximum empirical likelihood estimation (MELE), methods involving characteristic functions such as the integrated squared error estimator (ISEE) and stochastic programming techniques. These techniques appear in the various research papers such as Heathcote (1977), Sueishi and Nishiama (2005), Chan *et al.* (2009), Sant and Caruana (2015, 2012) and Sant and Caruana (2015). In the non-parametric setting, the techniques are subdivided in two groups: the low- and the high-frequency frameworks. In the latter, it is assumed that within each unit time interval, the number of observations n tends to infinity and at the same time, the number of time intervals N goes to infinity. On the contrary, in the low-frequency setting, n is fixed.

Rubin and Tucker (1959) pioneered the non-parametric estimation of infinitely divisible distributions within a high-frequency setting (although at the time such terminology was still not in use). In their paper, the authors proposed two non-parametric estimators for the function H. The first estimator of Rubin and Tucker (1959) can be defined as follows:

$$\hat{H}(u) = \frac{1}{N}\sum_{i=1}^{N}\sum_{j=1}^{n}\frac{X_{ij}^2}{1+X_{ij}^2}1_{\{X_{ij}\le u\}}, \qquad [11.4]$$

where X_{ij} represents the jth increment within the ith time interval. Other authors, which include Basawa and Brockwell (1982) and Gegler and Stadmuller (2010), also proposed estimators for the function H. The former only considered non-decreasing Lévy processes, and proposed three estimators for the function H. The second of which is identical to [11.4]. Moreover, these authors show that this estimator enjoys asymptotic normality. Gegler and Stadmuller (2010) apply the estimator of Rubin and Tucker only over the jump part of a Lévy process. As a result, their estimator cannot be defined over intervals close to and including 0.

Sant and Caruana (2017) proposed an estimator for the function G which is defined as follows:

$$\hat{G}(u) = \frac{1}{N}\sum_{i=1}^{N}\sum_{j=1}^{n}\frac{1}{n^{\frac{|X_{ij}|^\beta}{2}}}\frac{|X_{ij}|^{2-|X_{ij}|^\beta}}{1+|X_{ij}|^{2-|X_{ij}|^\beta}}1_{\{X_{ij}\le u\}}. \qquad [11.5]$$

The authors prove that [11.5] converges $\mathbb{P}$-almost surely to G at all points of continuity of the said function. Through a transformation of this estimator, which is discussed in the said paper, an estimator for the function H was also obtained.

Simulations revealed that this proposed estimator converged faster than the Rubin and Tucker estimator.

The non-parametric estimators discussed above all make use of discrete measures. However, they all have a significant disadvantage in that the points X_{ij} and the corresponding masses are not chosen in an optimal way. Indeed, the points X_{ij} simply correspond to the observed increments of an observed path of a Lévy Process. Hence, the main aim of this chapter is to propose an estimator of the measure Γ that can be defined in terms of G as follows:

$$\Gamma\left([a,b]\right) = G(b) - G(a), \qquad [11.6]$$

where $a, b \in \mathbb{R}$. This estimator will make use of the theory of discrete measures where the masses as well as their position are such that they minimize the objective function of a stochastic program which is described in the next section.

The rest of this chapter is organized as follows: in section 11.2, we will introduce the estimation strategy; in section 11.3, we will discuss some of the statistical properties of the estimator of Γ; section 11.4 presents some simulation results; and finally, section 11.5 contains some concluding remarks.

11.2. Estimation strategy

Let Ω be the set which contains all the possible paths of a Lévy process $(L_t)_{t\geq 0}$. Given a specific path $\omega \in \Omega$, let $\{X_{ij}(\omega)\}_{1\leq i\leq N, 1\leq j\leq n}$ denote a set of nN increments obtained from ω. As before, N denotes the number of time intervals, while n denotes the number of increments within each time interval. This double indexing of the increments is normal within the so-called high-frequency setting. Using these increments, we estimate the previously defined measure Γ that is associated with the distribution G. The estimator of Γ, which we denote by $\hat{\Gamma}$, is a random measure supported on a finite number of points, and is of the form

$$\hat{\Gamma} = \sum_{k=1}^{P} \hat{m}_k 1_{\hat{y}_k}, \qquad [11.7]$$

where $P \leq nN$ and $\hat{m}_k$ are the estimates of the masses m_k associated with the points y_k. The estimates of the latter are denoted by $\hat{y}_k$.

In the following, we first define a way to obtain both y_k and m_k through a stochastic program. Afterward, we estimate the position and the masses using the above-mentioned set of observed increments. Then, we discuss the convergence of $\hat{\Gamma}$ to Γ. In order to find the optimal points y_k, sometimes called atoms, as well as their respective masses m_k, we will primarily use theory related to the quantization of measures. The term *quantization* refers to the method of finding an optimal approximation of a probability density by a convex combination of a finite number of Dirac masses. The quality of such an approximation is measured in terms of the Wasserstein metric (sometimes called the Monge–Kantorovich metric).

Since G is non-decreasing and of bounded variation, then $G(\infty) = c$ for some $c \in \mathbb{R}^+$. Hence, the measure Γ/c associated with the distribution G/c belongs to the space of all probability measures, which is endowed with the Wasserstein metric. Throughout this chapter, we assume that to G/c, we can associate a density function g/c. Moreover, we assume that G/c has a finite second moment. This is the weakest assumption that we have to take and will be used below when using the Wasserstein metric. This assumption is not very restrictive and is satisfied by the well-known Lévy processes. Furthermore, more Lévy processes can be considered for which such an assumption is satisfied.

Since we have assumed that G/c has a finite second moment, then the optimal points y_k and their associated masses m_k, can be found by solving the following:

$$\inf \left[W \left(\sum_{k=1}^{P} \frac{m_k}{c} 1_{y_k}, \frac{g(x)}{c} dx \right) \middle| m_1, \ldots, m_P \geq 0, \sum_{k=1}^{P} m_k = c \right], \qquad [11.8]$$

where W denotes the Wasserstein metric.

From standard theory found in Graf and Luschgy (2000), Iacobelli (2015) and Caglioti *et al.* (2016), it was shown that the objective function in [11.8] can be written as follows:

$$\inf \left[W \left(\sum_{k=1}^{P} \frac{m_k}{c} 1_{y_k}, \frac{g(x)}{c} dx \right) \middle| m_1, \ldots, m_P \geq 0, \sum_{k=1}^{P} m_k = c \right]$$
$$= \mathbb{E} \left[\min_{1 \leq k \leq P} |y_k - x| \right]. \qquad [11.9]$$

In [11.9], we note that the right-hand side (RHS), i.e. $\mathbb{E} \left[\min_{1 \leq k \leq P} |y_k - x| \right]$, is indeed a stochastic program as discussed in various sources which include among others (Shinji 1962; Shapiro *et al.* 2009; Sueishi and Nishiama 2005). In stochastic programming, there are two main reformulations: the wait-and-see and the here-and-now. The stochastic program just defined in [11.9] belongs to the here-and-now reformulation. In the context of our problem, this stochastic program may be re-written as follows:

$$\mathbb{E} \left[\min_{1 \leq k \leq P} |y_i - x| \right] = \tfrac{1}{c} \int_{-\infty}^{\infty} \min_{1 \leq k \leq P} |y_k - x| g(x) dx. \qquad [11.10]$$

Moreover, in Theorem 7.5 in Graf and Luschgy (2000), it was shown that the optimal set of P points y_k that minimize the RHS of [11.9] has the property that as

$P \to \infty$ (in our case, as $n \to \infty$ or $N \to \infty$), the empirical measure generated from these points converges to

$$\frac{1}{P}\sum_{k=1}^{P} 1_{y_k} \to \frac{g^{1/(1+r)}}{\int_{-\infty}^{\infty} g(x)^{1/(1+r)}dx}. \qquad [11.11]$$

In this case, r has to satisfy the property that $\int_{-\infty}^{\infty} |z|^{r+1}\frac{g(z)}{c}dz < \infty$. Since we have assumed that G/c has finite second moment, we can take $r = 1$. By considering the RHS of [11.9], it can be shown from Iacobelli [9, pages 43 and 77], that

$$\tfrac{1}{c}\int_{\infty}^{\infty} \min_{1\leq k\leq P} |y_k - x|g(x)dx = \tfrac{1}{c}\sum_{k=1}^{P} \int_{y_{lk}}^{y_{uk}} |y_k - x|g(x)dx, \qquad [11.12]$$

where $y_{l1} = -\infty$, $y_{u1} = y_{l,2}$, $y_{uP} = \infty$, $y_{uk} = \frac{y_k+y_{k+1}}{2}$ and $y_{lk} = \frac{y_{k-1}+y_k}{2}$. From Iacobelli (2015), the optimal points y_k can be found through [11.12]. Through differentiation, this problem boils down to solve the following system of P equations involving P unknowns:

$$\frac{\partial \frac{1}{c}\sum\limits_{k=1}^{P}\int\limits_{y_{lk}}^{y_{uk}} |y_k - x|g(x)dx}{\partial y_k} = 0, \qquad [11.13]$$

for $k = 1, \ldots, P$. This can be re-written as follows:

$$2G(y_k) - G\left(\frac{y_k + y_{k-1}}{2}\right) - G\left(\frac{y_k + y_{k+1}}{2}\right) = 0, \qquad [11.14]$$

for $k = 1, \ldots, P$. Given the positions y_k, the best choice of the masses m_k that minimize [11.8] is explicit and is discussed in Lemmas 3.1 and 3.4 in Graf and Luschgy (2000). Adapting the results of these lemmas to the context of this chapter, we find that the value of m_k is given as follows:

$$m_k = \int_{y_{lk}}^{y_{uk}} g(z)dz = G(y_{uk}) - G(y_{lk}). \qquad [11.15]$$

Furthermore, since we have assumed that G/c has a finite second moment, as it was shown in Lemma 6.1 in Graf and Luschgy (2000) that given the points y_k and the corresponding weights m_k, the following holds:

$$\lim_{P\to\infty}\left(W\left(\sum_{k=1}^{P}\frac{m_k}{c}1_{y_k}, \frac{g(x)}{c}dx\right)\right) \to 0, \qquad [11.16]$$

provided that the corresponding probability distribution G is non-singular. This ensures that as the number of points $P \to \infty$, the approximation of the measure Γ through discrete measures approaches the actual measure Γ. However, since Γ is not known, we replace it in the above equations by an estimator defined in [11.7]. Furthermore, we replace G in [11.15] with its estimator $\hat{G}$, which was defined in [11.5]. This estimator was studied in Sant and Caruana (2017) and is defined through the previously defined increments X_{ij} obtained from an observed Lévy process. In the next section, we discuss the convergence of $\hat{y}_k$ and $\hat{m}_k$ as n, $N \to \infty$. Moreover, we show that $\hat{\Gamma}$ also converges almost surely to Γ.

11.3. Estimation of masses and the atoms

We start this section by considering the issue of estimating the optimal points y_k. In the previous section, we proposed to replace G by $\hat{G}$. As a result, [11.14] can be re-written as follows:

$$2\hat{G}(y_k) - \hat{G}\left(\frac{y_k + y_{k-1}}{2}\right) - \hat{G}\left(\frac{y_k + y_{k+1}}{2}\right) = 0, \qquad [11.17]$$

for $k = 1, \ldots, P$. Once the $\hat{y}_k$'s are computed, it is easy to estimate the masses. This can be done by replacing G, y_{lk} and y_{uk} by $\hat{G}$, $\hat{y}_{lk}$ and $\hat{y}_{uk}$, respectively, in [11.15]. Hence

$$\hat{m}_k = \hat{G}(\hat{y}_{uk}) - \hat{G}(\hat{y}_{lk}). \qquad [11.18]$$

The goal of this section is to show that

$$W\left(\sum_{k=1}^{P} \frac{\hat{m}_k}{c} 1_{\hat{y}_k}, \frac{g(x)}{c} dx\right) \qquad [11.19]$$

converges almost surely to 0. This step is not trivial and is presented in Theorem 11.4. However, in order to prove this theorem, we use a number of results presented in Theorems 11.1, 11.2 and 11.3. We start by defining the following discrete random measure:

$$H^*(S) = \sum_{j=1}^{P} \frac{m_k^*}{c} 1_{\hat{y}_k}(S), \qquad [11.20]$$

where $m_k^* = G(\hat{y}_{uk}) - G(\hat{y}_{lk})$ and $S \subseteq \mathbb{R}$. This random measure will be used in certain proofs below.

By the triangular inequality, we have that

$$W\left(\sum_{k=1}^{P}\frac{\hat{m}_k}{c}1_{\hat{y}_j},\frac{g(x)}{c}dx\right)\leq\underbrace{W\left(\sum_{k=1}^{P}\frac{\hat{m}_k}{c}1_{\hat{y}_k},\sum_{k=1}^{P}\frac{m_k^*}{c}1_{\hat{y}_k}\right)}_{1.1}$$
$$+\underbrace{W\left(\sum_{k=1}^{P}\frac{m_k^*}{c}1_{\hat{y}_k},\sum_{k=1}^{P}\frac{m_k}{c}1_{y_k}\right)}_{2.2}$$
$$+\underbrace{W\left(\sum_{k=1}^{P}\frac{m_k}{c}1_{y_k},\frac{g(x)}{c}dx\right)}_{3.3}. \qquad [11.21]$$

We observe that (3.3) in [11.21] has already been discussed in [11.16]. Hence, we proceed to consider (1.1) and (3.3) in [11.21]. However, before we consider these expressions, we present Theorem 11.1 below. This result will be frequently used in the following pages.

THEOREM 11.1.– *If $\hat{y}_{uk}$, $\hat{y}_{lk}$ are continuity points of G for $1 \leq k \leq P$, then $|\hat{m}_k - m_k^*|$ converges almost surely to* 0 *as* $n, N \to \infty$.

PROOF.– By the triangular inequality, we have that

$$\begin{aligned}|\hat{m}_k - m_k^*| &= \left|\left(\hat{G}(\hat{y}_{uk}) - \hat{G}(\hat{y}_{lk})\right) - (G(\hat{y}_{uk}) - G(\hat{y}_{lk}))\right|\\ &\leq \left|\hat{G}(\hat{y}_{uk}) - G(\hat{y}_{uk})\right| + \left|\hat{G}(\hat{y}_{lk}) - G(\hat{y}_{lk})\right|.\end{aligned}$$

From Sant and Caruana (2017), we know that both $\left|\hat{G}(\hat{y}_{uk}) - G(\hat{y}_{uk})\right|$ and $\left|\hat{G}(\hat{y}_{lk}) - G(\hat{y}_{lk})\right|$ converge almost surely to 0 since we know that $\hat{y}_{uk}$, $\hat{y}_{lk}$ are continuity points of G. □

Before we state Theorem 11.2, we recall that the Wasserstein distance between two discrete measures H^* and $\hat{H}$, both of which are defined on the atoms $\hat{y}_1, \ldots, \hat{y}_P$, can be expressed as follows:

$$W(H^*, \hat{H}) = \inf_q \sum_{ij} q_{ij}|y_i - y_j|, \qquad [11.22]$$

where q is the space of all possible matrices which satisfy the following conditions:

$$\sum_i q_{ij} = \frac{\hat{m}_j}{c} \text{ for each } j, \quad \text{and} \quad \sum_j q_{ij} = \frac{m_i}{c} \text{ for each } i,$$

where $q_{ij} \in [0,1] \times [0,1]$. For further details, we refer the interested reader to Nguyen (2011).

THEOREM 11.2.– *Given that $P \leq nN$ observations, if $\hat{y}_k$ is a point of continuity of G, then, $W\left(\sum_{j=1}^{P} \frac{\hat{m}_k}{c} 1_{\hat{y}_k}, \sum_{i=1}^{P} \frac{m_k^*}{c} 1_{\hat{y}_k}\right)$ converges almost surely to 0 as $n, N \to \infty$.*

PROOF.– We show that this result is true for the case when we have two masses and the case when we have three masses. We then move to the general case when we have P masses. The general case follows in a similar way to the previous two cases.

Case 1: we only have two masses

In this case, we have $\sum_{k=1}^{2} \frac{\hat{m}_k}{c} 1_{\hat{y}_k}, \sum_{k=1}^{2} \frac{m_k^*}{c} 1_{\hat{y}_k}$ and

$$W\left(\sum_{k=1}^{2} \frac{\hat{m}_k}{c} 1_{\hat{y}_k}, \sum_{k=1}^{2} \frac{m_k^*}{c} 1_{\hat{y}_k}\right) = \inf_q \left[(q_{12} + q_{21}) |\hat{y}_1 - \hat{y}_2|\right]. \qquad [11.23]$$

From the definition of q, it follows that we have the following constraints:

$$\begin{aligned} q_{11} + q_{12} &= m_1^*/c \\ q_{11} + q_{21} &= \hat{m}_1/c \end{aligned} \quad \text{and} \quad \begin{aligned} q_{21} + q_{22} &= 1 - (m_1^*/c) \\ q_{12} + q_{22} &= 1 - (\hat{m}_1/c) \end{aligned}$$

where $q_{ij} \in [0,1]$. Hence, to solve the above, we can use linear programming; thus, it can be re-written as follows:

$$\min\, (q_{12} + q_{21}) |\hat{y}_1 - \hat{y}_2|$$

subject to

$$q_{12} - q_{21} = (m_1^* - \hat{m}_1)/c$$

$$0 \leq q_{ij} \leq 1$$

The optimal solution to this problem is $|\hat{y}_1 - \hat{y}_2| \, |m_1^* - \hat{m}_1| / c$. Hence,

$$W\left(\sum_{k=1}^{2} \frac{\hat{m}_k}{c} 1_{\hat{y}_k}, \sum_{k=1}^{2} \frac{m_k^*}{c} 1_{\hat{y}_k}\right) = |\hat{m}_1 - m_1^*| \frac{|\hat{y}_1 - \hat{y}_2|}{c}. \qquad [11.24]$$

Moreover, we know, from Theorem 11.1 above, that $|\hat{m}_1 - m_1^*|$ converges almost surely to 0. Hence, the result follows.

Case 2: we have three masses

In this case, we have $\sum_{k=1}^{3} \frac{\hat{m}_k}{c} 1_{\hat{y}_k}, \sum_{k=1}^{3} \frac{m_k^*}{c} 1_{\hat{y}_k}$ and

$$\begin{aligned} W\left(\sum_{k=1}^{3} \frac{\hat{m}_k}{c} 1_{\hat{y}_k}, \sum_{k=1}^{3} \frac{m_k^*}{c} 1_{\hat{y}_k}\right) &= \inf_q \left[(q_{12}+q_{21})\left|\hat{y}_1-\hat{y}_2\right| + (q_{13}+q_{31})\left|\hat{y}_1-\hat{y}_3\right|\right. \\ &\quad \left. + (q_{32}+q_{23})\left|\hat{y}_2-\hat{y}_3\right|\right] \\ &\leq \inf_q \left[(q_{12}+q_{21})\rho_m + (q_{13}+q_{31})\rho_m\right. \\ &\quad \left. + (q_{32}+q_{23})\rho_m\right], \end{aligned}$$

where $\rho_m = \max\{|\hat{y}_1 - \hat{y}_2|, \hat{y}_1 - \hat{y}_3|, |\hat{y}_2 - \hat{y}_3|\} \geq 0$. The above inequality holds because $q_{ij} \geq 0 \; \forall i, j$. Moreover, the above elements of q must satisfy the following constraints:

$$\begin{aligned} q_{11} + q_{21} + q_{31} &= m_1^*/c & \qquad q_{11} + q_{12} + q_{13} &= \hat{m}_1/c \\ q_{12} + q_{22} + q_{32} &= m_2^*/c \quad \text{and} & \qquad q_{21} + q_{22} + q_{23} &= \hat{m}_2/c \\ q_{13} + q_{23} + q_{33} &= m_3^*/c & \qquad q_{31} + q_{32} + q_{33} &= \hat{m}_3/c \end{aligned}$$

Through the use of some algebraic manipulation, using the above six constraints, it can be shown that

$$\begin{aligned} (q_{12}+q_{21}) + (q_{13}+q_{31}) + (q_{32}+q_{23}) = \tfrac{1}{2}\big[&(m_1^*/c + \hat{m}_1/c - 2q_{11}) \\ &+ (m_2^*/c + \hat{m}_2/c - 2q_{22}) \\ &+ (m_3^*/c + \hat{m}_3/c - 2q_{33})\big] \end{aligned}$$

Moreover, q_{11} may be written as follows: $q_{11} = \lambda_1(\hat{m}_1/c) + (1-\lambda_1)(m_1^*/c)$, where $\lambda_1 = (cq_{11} - m_1^*)/(\hat{m}_1 - m_1^*)$.

Similarly, $q_{22} = \lambda_2(\hat{m}_2/c) + (1-\lambda_2)(m_2^*/c)$ and $q_{33} = \lambda_3(\hat{m}_3/c) + (1-\lambda_3)(m_3^*/c)$, for $\lambda_2, \lambda_3 \in \mathbb{R}$ which can be defined in a way similar to λ_1. Using these results, it follows that:

$$\begin{aligned} &W\left(\sum_{k=1}^{3} \frac{\hat{m}_k}{c} 1_{\hat{y}_k}, \sum_{k=1}^{3} \frac{m_k^*}{c} 1_{\hat{y}_k}\right) \\ &\leq \left(\frac{2\rho_m}{c}\left(\lambda_1\left|\hat{m}_1 - m_1^*\right| + \lambda_2\left|\hat{m}_2 - m_2^*\right| + \lambda_3\left|\hat{m}_3 - m_3^*\right|\right)\right). \end{aligned} \qquad [11.25]$$

As before, using Theorem 11.1, we have that $|\hat{m}_k - m_k^*|$ converges almost surely to 0. Hence, the result follows.

Case 3: we have P masses

To prove this case, we will simply generalize case 2 above. Let

$$W\left(\sum_{k=1}^{P}\frac{\hat{m}_k}{c}1_{\hat{y}_k},\sum_{j=1}^{P}\frac{m_k^*}{c}1_{\hat{y}_k}\right)=\inf_q\left[\sum_{i,j=1,i\neq j}^{P}2\left(q_{ij}+q_{ji}\right)\rho(\hat{y}_i,\hat{y}_j)\right]$$

$$\leq \inf_q\left[\rho_m\sum_{i,j=1,i\neq j}^{P}2\left(q_{ij}+q_{ji}\right)\right],$$

where $\rho_m=\max\{\rho(\hat{y}_i,\hat{y}_j),0\leq i,j\leq P,i\neq j\}\geq 0$.

Note that the above inequality holds because $q_{ij}\geq 0\ \forall i,j$. Moreover, the above elements of q must satisfy the following constraints:

$$\sum_{i=1}^{P}q_{ij}=m_j^*/c\ \text{ for each } j,\qquad \text{and}\qquad \sum_{j=1}^{P}q_{ij}=\hat{m}_i/c\ \text{ for each } i,$$

where $1\leq i,j\leq P$. Using these conditions, it can be shown that

$$\sum_{i,j=1,i\neq j}^{P}2\left(q_{ij}+q_{ji}\right)=\sum_{k=1}^{P}\left(\frac{m_k^*}{c}+\frac{\hat{m}_k}{c}-2q_{kk}\right). \qquad [11.26]$$

As before, we have that

$$W\left(\sum_{k=1}^{P}\frac{\hat{m}_k}{c}1_{y_k},\sum_{k=1}^{P}\frac{m_k^*}{c}1_{y_k}\right)\leq\left(\frac{2\rho_m}{c}\sum_{k=1}^{P}\left(\lambda_k\left|\hat{m}_k-m_k^*\right|\right)\right). \qquad [11.27]$$

From Theorem 11.1, we know that $|\hat{m}_k-m_k^*|$ converges almost surely to 0 for all k as $n,N\to\infty$. Moreover, we have $\sum_{k=1}^{P}\left(|\hat{m}_k-m_k^*|\right)$ converges almost surely to 0 as $n,N\to\infty$. Hence, $W\left(\sum_{k=1}^{P}\hat{m}_k1_{\hat{y}_k},\sum_{k=1}^{P}m_k^*1_{\hat{y}_k}\right)$ also converges almost surely to 0 as $n,N\to\infty$. □

We next consider the term (2.2) in [11.21], i.e. $W\left(\sum_{k=1}^{P}\hat{m}_k1_{\hat{y}_k},\sum_{k=1}^{P}m_k^*1_{\hat{y}_k}\right)$.

In this case, we note that the position of the masses and the masses themselves are different, unlike in the previous case. Nevertheless, the following expression still holds:

$$W\left(\sum_{k=1}^{P}\frac{m_k^*}{c}1_{\hat{y}_k},\sum_{k=1}^{P}\frac{m_k}{c}1_{y_k}\right)=\inf_q\sum_{i=1}^{P}\sum_{j=1}^{P}q_{ij}\left|\hat{y}_i-y_j\right|, \qquad [11.28]$$

subject to the constraints

$$\sum_{i=1}^{P} q_{ij} = \frac{m_j}{c} \text{ for each } j, \quad \text{and} \quad \sum_{j=1}^{P} q_{ij} = \frac{m_i^*}{c} \text{ for each } i,$$

where $1 \leq i, j \leq P$.

THEOREM 11.3.– *If $\hat{y}_{uk}$ and $\hat{y}_{lk}$ are the points of continuity of G, then we have that*

$$W\left(\sum_{k=1}^{P} \frac{m_k^*}{c} 1_{\hat{y}_k}, \sum_{k=1}^{P} \frac{m_k}{c} 1_{y_k}\right) \qquad [11.29]$$

converges almost surely to 0 as $n, N \to \infty$.

PROOF.– Following arguments similar to those in Theorem 11.2, we have that

$$W\left(\sum_{k=1}^{P} \frac{m_k^*}{c} 1_{\hat{y}_k}, \sum_{k=1}^{P} \frac{m_k}{c} 1_{y_k}\right) \leq \frac{1}{c} \sum_{k=1}^{P} |\hat{y}_k - y_k| m_k^* + \frac{\rho_m}{c} \sum_{k=1}^{P} (|m_k^* - m_k|),$$

[11.30]

whereas before $\rho_m = \max\{|y_i - y_j|, 1 \leq i, j \leq P, i \neq j\} \geq 0$.

Moreover,

$$|m_k^* - m_k| = |(G(\hat{y}_{uk}) - G(\hat{y}_{lk})) - (G(y_{uk}) - G(y_{lk}))|$$
$$\leq |G(y_{uk}) - G(\hat{y}_{uk})| + |G(y_{lk}) - G(\hat{y}_{lk})|,$$

Since the derivative of G/c is a density function, then by definition, it must be bounded and thus G is Lipchitz. This implies that the following inequalities hold:

$$|G(y_{uk}) - G(\hat{y}_{uk})| \leq \gamma |y_{uk} - \hat{y}_{uk}|, \text{ and } |G(y_{lk}) - G(\hat{y}_{lk})| \leq \sigma |y_{uk} - \hat{y}_{uk}|,$$

for some finite constants γ and σ. Moreover, $|\hat{y}_{uk} - y_{uk}|$ and $|\hat{y}_{lk} - y_{lk}|$ both go to zero almost surely as $\hat{G}$ goes to G almost surely at continuity points $\hat{y}_{uk}$ and $\hat{y}_{lk}$. This implies that $W\left(\sum_{k=1}^{P} \frac{m_k^*}{c} 1_{\hat{y}_k}, \sum_{k=1}^{P} \frac{m_k}{c} 1_{y_k}\right)$ converges almost surely to 0 as $n, N \to \infty$. □

THEOREM 11.4.– *If $\hat{y}_{uk}$ and $\hat{y}_{lk}$ are the points of continuity of G, then we have that*

$$W\left(\sum_{k=1}^{P} \frac{\hat{m}_k}{c} 1_{\hat{y}_k}, \frac{g(x)}{c} dx\right) \qquad [11.31]$$

converges almost surely to 0 *as $P, n, N \to \infty$.*

PROOF.– To prove this result, we just have to combine Theorem 11.2 and Theorem 11.3 with [11.21] and [11.16]. □

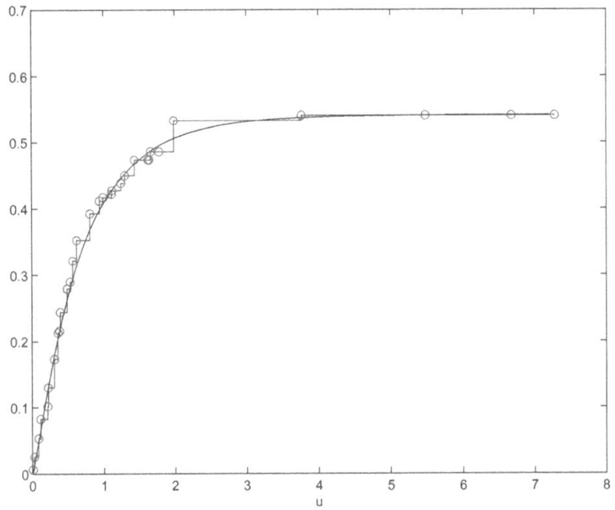

Figure 11.1. *Comparison of* $\Gamma[-\infty, u]$ *and* $\hat{\Gamma}[-\infty, u]$ *for the gamma process. For a color version of this figure, see www.iste.co.uk/dimotikalis/analysis2.zip*

11.4. Simulation results

In this section, we briefly discuss the results obtained from a number of simulations, using different Lévy processes to compare the proposed estimator $\hat{\Gamma}$ with Γ. The chosen Lévy processes are: the gamma process and the Cauchy process. We note that these two processes are pure jump Lévy processes.

In both simulations, the position of a total of 31 atoms was estimated. Moreover, the weights associated with these atoms were also computed. The number of positions taken was reduced to 31 so as to easily compare visually the actual measure with its estimate. If more points were included, it would not be possible to compare visually both measures as they would overlap. We observe that in both diagrams, the masses were not located at random along the x-axis but rather at key points. Indeed, we observe that few masses appear in areas where the curve is relatively flat as opposed to other areas. Clearly, as the number of masses increases, the estimates would get progressively better.

For the gamma process, we chose the shape and scale parameters to be both equal to 1. In both cases, the number N of unit time intervals was chosen to be equal to 10. Moreover, in each interval, we had a total of 1000 observations.

In Figure 11.1, we compare $\Gamma[0, u]$ and $\hat{\Gamma}[0, u]$. We can observe that with just 31 atoms, $\hat{\Gamma}[0, u]$ is a good estimator of $\Gamma[0, u]$. We next consider the Cauchy process with location and scale parameters being equal to 1 and 0.05, respectively. In this simulation, we took the same number of unit time intervals and the same number of observations within each interval. In Figure 11.2, we compare $\Gamma[-\infty, u]$ and $\hat{\Gamma}[-\infty, u]$. As in the previous simulation, we observe that with just 31 atoms, $\hat{\Gamma}[-\infty, u]$ is a good estimator of $\Gamma[-\infty, u]$.

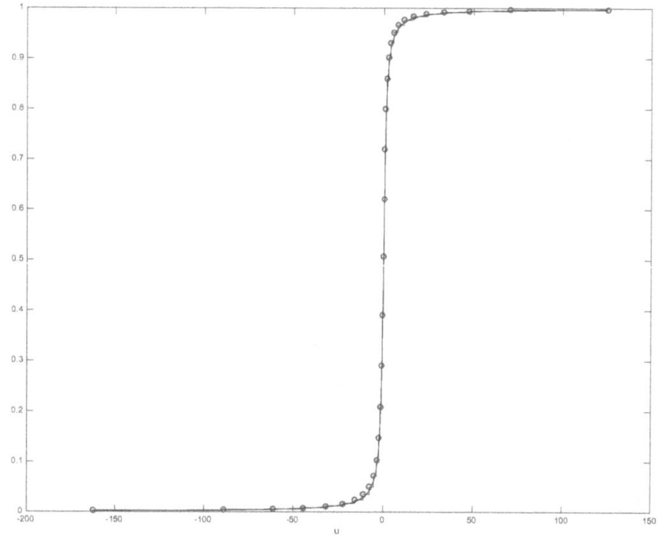

Figure 11.2. *Comparison of $\Gamma[-\infty, u]$ and $\hat{\Gamma}[-\infty, u]$ for the Cauchy process. For a color version of this figure, see www.iste.co.uk/dimotikalis/analysis2.zip*

11.5. Conclusion

In this chapter, we applied the theory of quantization of measures and proposed an estimator for the measure associated with the function G. The estimator is defined as a sum of discrete measures where the masses and their respective positions are such that these minimize the Wasserstein distance between the measure Γ and the proposed estimator. We have also shown that as the number of masses tends to infinity, the Wasserstein distance between the measure and the estimator tends to zero almost surely. Ideally, we should also obtain confidence intervals around the proposed estimator; however, these will appear in a future work.

Simulations have shown that with just 31 points, good estimates were obtained for the gamma process and for the Cauchy process. Moreover, the diagrams reveal that the points are not evenly spaced out. Indeed, there are a few masses where the curve is relatively flat. This is in sharp contrast with other areas. The same cannot be said about the other estimators, such as the Rubin and Tucker estimator discussed in Rubin and Tucker (1959) and its variants discussed in Sant and Caruana (2017), where the position of the masses simply coincided with the size of the increments obtained from an observed Lévy process.

11.6. References

Applebaum, D. (2004). *Lévy Processes and Stochastic Calculus*. Cambridge University Press.

Basawa, I. and Brockwell, P. (1982). Non-parametric estimation for non-decreasing Lévy processes. *Journal of the Royal Statistical Society, Series B (Methodological)*, 44, 262–269.

Bertoin, J. (1996). *Lévy Processes*. Cambridge University Press.

Caglioti, E., Golse, F., Iacobelli, M. (2016). Quantization of measures and gradient flows: A perturbative approach in the 2-dimensional case. *HAL*.

Chan, N.H., Chen, S., Peng, L., Yu, C.L. (2009). Empirical likelihood methods based on characteristic functions with applications to Lévy processes. *Journal of the Royal Statistical Society. Series B (Methodological)*, 104(448), 1612–1630.

Gegler, A. and Stadmuller, U. (2010). Estimation of the characteristics of a Lévy process. *Journal of Statistical Planning and Inference*, 140, 1481–1496.

Graf, S. and Luschgy, H. (2000). *Foundations of Quantization for Probability Distributions*. Springer-Verlag, Berlin, Heidelberg.

Heathcote, C. (1977). The integrated squared error estimation of parameters. *Biometrika*, 64, 255–264.

Iacobelli, M. (2015). Dynamics of large particle systems. PhD Thesis, University of Rome.

Kyprianou, A.E. (2006). *Introductory Lectures on Fluctuations of Lévy Processes with Applications*. Springer, Berlin.

Nguyen, X. (2011). Wasserstein distance for discrete measures and convergence in nonparametric mixture models. Technical Report 527, University of Michigan.

Rubin, H. and Tucker, H.G. (1959). Estimating the parameters of a differential process. *Ann. Math. Statist.*, 30, 641–658.

Sant, L. and Caruana, M.A. (2012). Products of characteristic functions in Lévy processes parameter estimation. *SMTDA Conference Proceedings*, Crete.

Sant, L. and Caruana, M.A. (2015a). Estimation of Lévy processes through stochastic programming. In *Stochastic Modelling, Data Analysis and Statistical Applications*, 1st edition, Filus, L., Oliviera, T., Skiadas, C.H. (eds). ISAST.

Sant, L. and Caruana, M.A. (2015b). Incorporating the stochastic process setup in parameter estimation. *Methodology and Computing in Applied Probability*, 17(4), 1029–1037.

Sant, L. and Caruana, M.A. (2017). Choosing tuning instruments for generalized Rubin-Tucker Lévy measure estimators. *17th ASMDA Conference Proceedings*, London.

Sato K. (1999). *Lévy Processes and Infinitely Divisible Distribtuions*. Cambridge University Press.

Shapiro, A., Dentcheva, D., Ruszcynski, A. (2009). *Lectures on Stochastic Programming*. MPS-SIAM, University City, Philadelphia.

Shinji, K. (1962). On stochastic programming and its application to production horizon problem. *Hitotshubashi Journal of Arts and Sciences*, 2(1), 23–55.

Sueishi, N. and Nishiama, Y. (2005). Estimation of Lévy processes in mathematical finance: A comprehensive study. *MODSIM 2005 International Congress on Modelling and Simulation*, 953–959.

12

A Flexible Mixture Regression Model for Bounded Multivariate Responses

Compositional data are defined as vectors with strictly positive elements subject to a unit-sum constraint. The aim of this contribution is to propose a regression model for multivariate continuous variables with bounded support by taking into consideration the flexible Dirichlet (FD) distribution that can be interpreted as a special mixture of Dirichlet distributions. The FD distribution is an extension of the Dirichlet one, which is contained as an inner point, and it enables a greater variety of density shapes in terms of tail behavior, asymmetry and multimodality. A convenient parameterization of the FD is provided which is variation independent and facilitates the interpretation of the mean vector of each mixture component as a piecewise increasing linear function of the overall mean vector. A multivariate logit strategy is adopted to regress the vector of means, which is itself constrained to add up to one, onto a vector of covariates. Intensive simulation studies are performed to evaluate the fit of the proposed regression model, particularly in comparison with the Dirichlet regression model. Inferential issues are dealt with by a (Bayesian) Hamiltonian Monte Carlo algorithm.

12.1. Introduction

Compositional data, namely proportions of some whole, are defined on the simplex space $\mathcal{S}^D = \{\mathbf{Y} : Y_j > 0, j = 1, \dots, D, \sum_{j=1}^{D} Y_j = 1\}$. Many fields of research, such as geology, psychology, economics to cite but a few, feature multivariate bounded responses on the simplex. A traditional approach to deal with compositional data consists of transforming them, for example, through log-ratios, in order to recover standard methods based on the normality-assumption (see the pioneer work of

Chapter written by Agnese M. DI BRISCO and Sonia MIGLIORATI.

(Aitchison 1986)). Nevertheless, this strategy has proved to be fallacious in case of violation of the hypothesis of homoscedasticity and in the presence of skewness, which are quite common issues for bounded data. Furthermore, it makes it difficult to interpret the estimated parameters with respect to the original multivariate response.

An alternative approach consists of modeling compositional data directly on the simplex, taking advantage of proper distributions on this bounded space such as the Dirichlet distribution. Regression models for compositional data based on the Dirichlet distribution (Campbell and Mosimann 1987; Hijazi 2003) prove to behave satisfactorily in many applications (Gueorguieva *et al.* 2008; Maier 2014). However, the structure of the Dirichlet distribution makes it unsuitable to cope with relevant density shapes such as multi-modality and heavy tails, possibly induced by outlying observations. In this regard, many other distributions on the simplex have been proposed as alternatives to the Dirichlet, among which the flexible Dirichlet (FD) distribution that has the peculiarity of being a special mixture of Dirichlet distributions with interesting theoretical properties (Ongaro and Migliorati 2013; Migliorati *et al.* 2016). The aim of this work is to define a regression model based on the FD distribution, referred to as the flexible Dirichlet regression (FDReg) model (a first definition and a simple illustration can be found in Di Brisco and Migliorati (2017)). Moreover, we aim to extend the comprehension of the potentialities of the FDReg model and its fit capacity compared to the Dirichlet regression (DirReg) model through simulation studies under several scenarios. Note that a previous work on the univariate version of the FD regression has already shown its remarkable fitting capacity for a variety of data patterns (Migliorati *et al.* 2018).

The rest of this chapter is organized as follows. Section 12.2 introduces the Dirichlet and the FD distributions. Moreover, it shows a convenient parameterization of the latter for regression purposes, and it describes the FDReg model for compositional data. Section 12.3 provides details on a Bayesian approach to inference suitable for the FDReg model. Section 12.4 illustrates several simulation studies that have been performed to evaluate the behavior of the proposed regression model. Finally, section 12.5 is devoted to our final comments.

12.2. Flexible Dirichlet regression model

Let us define a vector $\mathbf{Y} = (Y_1, \ldots, Y_D)'$ following a Dirichlet distribution, i.e. $\mathbf{Y} \sim \mathcal{D}(\boldsymbol{\alpha})$ with a density function (df) equal to:

$$f_D(\mathbf{y}; \alpha_1, \ldots, \alpha_D) = \frac{1}{B(\boldsymbol{\alpha})} \prod_{j=1}^{D} y_j^{\alpha_j - 1}, \qquad [12.1]$$

where $\mathbf{y} \in \mathcal{S}^D$, $B(\boldsymbol{\alpha}) = \prod_{j=1}^{D} \Gamma(\alpha_j)/\Gamma(\sum_{j=1}^{D} \alpha_j)$ and $\alpha_1, \ldots, \alpha_D > 0$. In a regression perspective, the following mean-precision parameterization proves to be convenient:

$$\begin{cases} \alpha^+ = \sum_{j=1}^{D} \alpha_j, \\ \bar{\alpha}_j = \frac{\alpha_j}{\alpha^+} \quad j = 1, \ldots, D, \end{cases} \qquad [12.2]$$

with $\sum_{j=1}^{D} \bar{\alpha}_j = 1$.

An interesting generalization of the Dirichlet, referred to as FD, has been proposed to greatly extend the density shapes of the former and its flexibility in terms of dependence/independence structure while preserving many of its theoretical properties (Ongaro and Migliorati 2013). The FD distribution is a special mixture of D Dirichlet distributions. Given a vector $(Y_1, \ldots, Y_D)'$ following an FD distribution, $\mathbf{Y} \sim FD(\boldsymbol{\alpha}, \mathbf{p}, \tau)$, its df is equal to:

$$f_{FD}(\mathbf{y}; \alpha_1, \ldots, \alpha_D, p_1, \ldots, p_D, \tau) = \sum_{h=1}^{D} p_h f_D(\mathbf{y}, \boldsymbol{\alpha} + \tau \mathbf{e}_h), \qquad [12.3]$$

where $h = 1 \ldots, D$ denotes the component of the mixture, $\alpha_1, \ldots, \alpha_D > 0$, $\tau > 0$, $p_1, \ldots, p_D > 0$, $\sum_{h=1}^{D} p_h = 1$ and $\mathbf{e}_h$ is a vector of zeros except for position h–th that is equal to 1.

A new parameterization of the FD distribution that highlights the mean vector of $\mathbf{Y}$ is derived. First, please note that each component $\mathbf{Y}_h$ of the mixture has a mean vector equal to:

$$\boldsymbol{\lambda}_h = \mathbb{E}(\mathbf{Y}_h) = \frac{\boldsymbol{\alpha}}{\alpha^+ + \tau} + \frac{\tau}{\alpha^+ + \tau} \mathbf{e}_h. \qquad [12.4]$$

Equation [12.4] makes clear the peculiarity of the mixture structure of the FD distribution. In fact, it ensures that each component of the mixture is distinguishable – the h-th element of the mean vector of the h-th component being higher than the corresponding element of the mean vector of the remaining components – thus avoiding the label switching problem (Frühwirth-Schnatter 2006). Moreover, within the FD, $\alpha^+ = \sum_{j=1}^{D} \alpha_j$ plays the role of a precision parameter and is the same for all the components of the mixture.

Let us define $w = \frac{\tau}{\alpha^+ + \tau}$, then the mean of the j–th component of vector $\mathbf{Y}$ is:

$$\mu_j = \mathbb{E}(Y_j) = \sum_{h=1}^{D} p_h \lambda_{hj} = \bar{\alpha}_j (1 - w) + p_j w, \qquad [12.5]$$

where $\bar{\alpha}_j = \frac{\alpha_j}{\alpha^+}$ for $j = 1, \ldots, D$. It is worth noting that $0 < \bar{\alpha}_j < 1$, from which, after some algebra, it follows that $0 < w < \min\left\{1, \min_j\left\{\frac{\mu_j}{p_j}\right\}\right\}$. Thus, the normalized version of w, denoted by $0 < w^* < 1$, takes the form:

$$w^* = \frac{w}{\min\left\{1, \min_j\left\{\frac{\mu_j}{p_j}\right\}\right\}}.$$

The new parameterization of the FD distribution depending on $\boldsymbol{\mu} = (\mu_1, \ldots, \mu_D)' \in \mathcal{S}^D$, $\mathbf{p} \in \mathcal{S}^D$, $\alpha^+ > 0$, and $0 < w^* < 1$ is variation independent, meaning that no constraint exists among the parameter spaces. This has a positive impact particularly on Bayesian inferential aspects (see section 12.3 for details). Moreover, such new parameterization determines a clear interpretation of the mean vector of each mixture component [12.4] as a piecewise increasing linear function of the overall mean vector, which can be seen by combining [12.4] and [12.5].

Let us now focus on regression issues. To such an end, let us consider a response vector $\mathbf{y}_i = (y_{i1}, \ldots, y_{iD})'$ on the simplex and a corresponding vector of covariates $\mathbf{x}_i = (x_{i0}, x_{i1}, \ldots, x_{ik})'$ for subject $i = 1, \ldots, n$. Furthermore, let us assume that $\mathbf{Y}_i$ is FD distributed and that we aim to regress its mean vector onto covariates. Since $\boldsymbol{\mu}_i$ belongs to the simplex too, a GLM-type regression model (McCullagh and Nelder 1989) for the mean has to take into account the constraints of positivity and unit-sum. In this regard, we take advantage of a multinomial logit strategy defining:

$$\log\left(\frac{\mu_{ij}}{\mu_{iD}}\right) = \mathbf{x}_i'\boldsymbol{\beta}_j, \quad j = 1, \ldots, D, \qquad [12.6]$$

from which it follows that:

$$\mu_{ij} = \frac{\exp\left(\mathbf{x}_i'\boldsymbol{\beta}_j\right)}{\sum_{j=1}^{D}\exp\left(\mathbf{x}_i'\boldsymbol{\beta}_j\right)}, \quad j = 1, \ldots, D, \qquad [12.7]$$

where $\boldsymbol{\beta}_j = (\beta_{j0}, \beta_{j1}, \ldots, \beta_{jk})'$ is a vector of regression coefficients. Please note that by construction $\boldsymbol{\beta}_D = \mathbf{0}$.

Under the assumption that $\mathbf{Y}_i$ follows a Dirichlet distribution, for $i = 1, \ldots, n$, the DirReg model is recovered by replacing μ_{ij} with $\bar{\alpha}_{ij}$ into equations [12.6] and [12.7].

12.3. Inferential issues

A Bayesian approach to inference based on Markov Chain Monte Carlo (MCMC) is particularly suitable to cope with complex models with many parameters and hierarchical models, such as mixtures (Gelman *et al.* 2014). A likelihood-based inference would require cumbersome integration and optimization; instead, a Bayesian approach to inference has proven to be computationally tractable. A recent solution to simulate posterior distributions for the parameter vector is the Hamiltonian Monte Carlo (HMC) algorithm (Duane *et al.* 1987; Neal 1994), a generalization of the Metropolis algorithm which combines MCMC and deterministic simulation methods.

The HMC is implemented in Stan modeling language using the standard No-U-Turn Sampler (NUTS) (Gelman *et al.* 2014; Stan Development Team 2016). To make inference on the samples from the posterior distributions, it is required to specify the full likelihood function and prior distributions for the unknown parameters. Given a sample of i.i.d response vectors $\mathbf{y}_i$ $(i = 1, \ldots, n)$, the likelihood function of the FDReg model is equal to:

$$L(\boldsymbol{\eta}|\mathbf{y}) = \prod_{i=1}^{n} f^*_{FD}(\mathbf{y}_i; \boldsymbol{\beta}_1, \ldots, \boldsymbol{\beta}_{D-1}, \alpha^+, \mathbf{p}, w^*),$$

where $f^*_{FD}(\cdot)$ is the df of the FD distribution under the new parameterization depending on the vector of unknown parameters $\boldsymbol{\eta} = (\boldsymbol{\beta}_1, \ldots, \boldsymbol{\beta}_{D-1}, \alpha^+, \mathbf{p}, w^*)'$. With respect to the prior choice, we favor non- or weakly informative priors with the purpose of inducing a minimum impact on the posteriors. Since the parametric space is variation independent, we might further assume prior independence.

The chosen prior for regression parameter $\boldsymbol{\beta}_j$ is (a diffuse) multivariate normal with zero mean vector and a diagonal covariance matrix with "large" values for the variances to induce flatness, i.e. non-informativeness. For the remaining parameters of the model, we assign a $gamma(g, g)$ prior with $g = 0.001$ to the precision parameter α^+, a Laplace prior $Unif(0, 1)$ to w^* and a non-informative Dirichlet with hyperparameter $\mathbf{1}$ to $\mathbf{p}$.

12.4. Simulation studies

We investigate the performance of the FDReg and DirReg models by setting up some simulation studies. We adopt a Bayesian approach to inference, as described in section 12.3, for both models in all studies. In particular, we run chains of a length of 10,000, having discarded the first half, and we check the convergence to the target distribution by adopting graphical tools, such as trace-plots, density-plots and autocorrelation-plots, as well as diagnostic measures among which are the potential scale reduction factor, the effective sample size, and the Raftery–Lewis test (Gelman *et al.* 2014). Monte Carlo posterior means and 95% credible intervals (CIs) are obtained by replicating each study 500 times and taking the average.

12.4.1. *Simulation study 1: presence of outliers*

As a first baseline scenario (i), we consider a data generating process that follows a DirReg model. The sample size is $n = 250$, and the mean vector of the Dirichlet distributed multivariate response with D=3 is regressed (see equation [12.7]) onto a quantitative covariate x uniformly distributed in $(-0.5, 0.5)$. Regression coefficients are set equal to $\beta_{10} = 1$, $\beta_{11} = 2$, $\beta_{20} = 0.5$, $\beta_{21} = -3$, and the precision parameter is $\alpha^+ = 50$. From Table 12.1, it emerges that both models provide accurate estimates of regression coefficients and of the precision parameter. Thus, the fitted regression lines of the DirReg and FDReg models are almost entirely overlapping. Furthermore, it is worth noting that the true values of all parameters are included in the CIs of both models. The FDReg model fits the Dirichlet structure of data very well by estimating three equally weighted mixture components, i.e. $p_j \approx 1/3$ for $j = 1, 2, 3$; moreover, the estimate of the parameter w^*, which measures the distance between component means, is near zero.

A fully Bayesian criterion that balances between the goodness of fit of a model and its complexity and that is properly defined for mixture models is the widely applicable information criterion (WAIC) (the lower the better) (Vehtari *et al.* 2017). Please note that WAIC values, obtained as averages over the 500 replications, are similar among the models (see Table 12.1), thus suggesting that they both well-adapt to data, despite the fact that the Dirichlet is the favored one being the "true" data generating model.

Scenario (i)	FDReg		DirReg	
	Mean	CI	Mean	CI
$\beta_{10} = 1$	1.000	(0.949;1.052)	0.999	(0.949;1.051)
$\beta_{20} = 0.5$	0.502	(0.446;0.557)	0.502	(0.446;0.557)
$\beta_{11} = 2$	2.000	(1.810;2.190)	1.999	(1.809;2.187)
$\beta_{21} = -3$	-2.978	(-3.181;-2.777)	-2.977	(-3.179;-2.772)
$\alpha^+ = 50$	49.922	(44.253;56.671)	48.932	(43.512;55.407)
p_1	0.331	(0.297;0.366)		
p_2	0.334	(0.300;0.376)		
p_3	0.334	(0.301;0.362)		
w^*	0.037	(0.031;0.045)		
WAIC	-1583.702		-1583.648	

Table 12.1. *Posterior means and CIs of unknown parameters together with WAIC based on 500 replications for the FDReg and DirReg models*

These simulation results suggest that the FDReg model, despite guaranteeing a greater flexibility and a richer parameterization than the DirReg model, is capable of also accommodating for simpler scenarios without the risk of over-fitting and penalization due to its higher number of parameters.

Next, we contaminate the baseline scenario (i) so as to induce outliers. Thus, we sample a subset of 25 (10%) response values and we transform them into artificial outliers by taking advantage of the perturbation operation on the simplex (Pawlowsky-Glahn *et al.* 2015), which can be seen as analogous to addition onto real space. Indeed, a simple shift operation on randomly selected response values is unsuitable since it might generate values outside the support. Let us consider two vectors, $\mathbf{y}$ and $\boldsymbol{\delta}$, both defined on the simplex and assuming the role of the perturbed and perturbing elements. The perturbation operation is thus defined as follows:

$$\mathbf{y} \oplus \boldsymbol{\delta} = \mathcal{C}\{y_1 \cdot \delta_1, \ldots, y_D \cdot \delta_D\} \in \mathcal{S}^D, \qquad [12.8]$$

where $\mathcal{C}\{\cdot\}$ is the closure operation, $\mathcal{C}\{\mathbf{q}\} = \left\{ \frac{q_1}{\sum_{j=1}^{D} q_j}, \ldots, \frac{q_D}{\sum_{j=1}^{D} q_j} \right\}$, with $q_j > 0, \forall j = 1, \ldots, D$. The vector resulting from perturbation operation in [12.8] lies on the simplex as well. The neutral element of the perturbation operation is $(1/D, \ldots, 1/D)^{'}$ and so if element y_j is perturbed by a δ_j greater (lower) than $1/D$, the perturbation is upward (downward).

The presence of outlying observations is challenging in statistical modeling. Different from univariate outliers, multivariate outliers are hard to inspect graphically and they might not be extreme along one single coordinate (Filzmoser and Hron 2008). The purpose of this simulation study is, thus, to investigate if the mixture structure of the FDReg, despite not being specifically designed with this aim, is capable of handling some patterns of outliers.

In particular, we evaluate three scenarios of perturbation by setting the perturbing factor $\boldsymbol{\delta}$ to (ii) $(0.8, 0.1, 0.1)^{'}$, (iii) $(0.1, 0.8, 0.1)^{'}$ and (iv) $(0.1, 0.1, 0.8)^{'}$. Figure 12.1 shows the ternary plots (i.e. proper graphical representations for compositional data when D=3) of one simulated sample at baseline (i) and in the three scenarios of perturbation. It is worth noting that the perturbed response values (light blue points) are far from the central cloud of observations only in scenario (iv), and they are extreme values only for the third coordinate. Differently, in scenarios (ii) and (iii), the perturbed values are shifted towards the top-right and bottom-left corners of the ternary diagram, respectively, but they are not far away from the central body of observations. Interestingly, in scenario (ii), the cluster of perturbed values turns out to be influential with respect to the first marginal coordinate but not to the log-ratio between the first and last coordinates. The same holds in scenario (iii) for the second marginal coordinate.

In all scenarios (see Tables 12.2, 12.3 and 12.4), the FDReg model provides a better fit (lower WAIC value) than the DirReg. Nevertheless, neither model produces robust estimates of the regression coefficients of the mean vector. For example, in scenario (iii), the 25 randomly selected y_{i2} are perturbed upward and, as a result, both the DirReg and the FDReg regression curves are flattened, with estimated regression coefficients $\hat{\beta}_{11}$ and $\hat{\beta}_{21}$ closer to zero than the true value. Therefore, it is necessary to deepen the analysis to understand the reason why the FDReg model provides a better fit than the DirReg model despite not determining an increase in point estimate robustness. It is worth noting that the mixture structure of the FDReg model provides the required flexibility to cluster the response values into outlying and not-outlying values. Indeed, in all scenarios, one component of the mixture is dedicated to describing the majority of observations, about 90%, another component is dedicated to the 10% of outlying values and the remaining component is dedicated to a residual amount of less than 1% observations. Parameter w^*, which measures the distance between the component means, is high, about 0.6, in all scenarios.

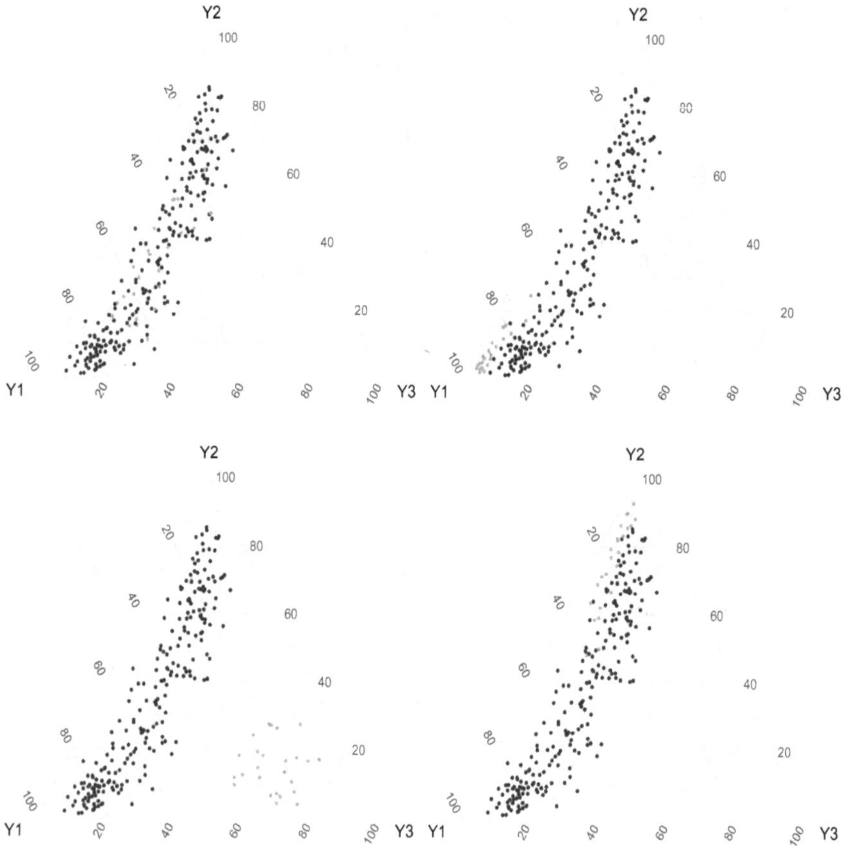

Figure 12.1. *Clockwise from top-left panel: ternary plots at baseline (i) and in case of perturbations (ii), (iii) and (iv) of one simulated sample. The perturbed response values are in light blue. For a color version of this figure, see www.iste.co.uk/dimotikalis/analysis2.zip*

Scenario (ii)	FDReg		DirReg	
	Mean	CI	Mean	CI
$\beta_{10} = 1$	1.230	(1.186;1.277)	1.149	(1.106;1.195)
$\beta_{20} = 0.5$	0.578	(0.522;0.630)	0.465	(0.414;0.509)
$\beta_{11} = 2$	1.653	(1.459;1.824)	1.828	(1.648;2.001)
$\beta_{21} = -3$	-2.632	(-2.815;-2.443)	-2.907	(-3.079;-2.728)
α^+	22.180	(20.038;24.376)	17.880	(16.373;19.387)
p_1	0.128	(0.115;0.143)		
p_2	0.867	(0.852;0.880)		
p_3	0.005	(0.004;0.005)		
w^*	0.672	(0.631;0.707)		
WAIC	-1291.953		-1143.081	

Table 12.2. *Posterior means and CIs of unknown parameters together with WAIC based on 500 replications for the FDReg and DirReg models. Perturbation factor* $(0.8, 0.1, 0.1)'$

Scenario (iii)	FDReg		DirReg	
	Mean	CI	Mean	CI
$\beta_{10} = 1$	1.025	(0.977;1.075)	0.919	(0.874;0.967)
$\beta_{20} = 0.5$	0.816	(0.756;0.874)	0.690	(0.640;0.738)
$\beta_{11} = 2$	1.825	(1.633;2.015)	1.959	(1.793;2.132)
$\beta_{21} = -3$	-2.233	(-2.438;-2.030)	-2.479	(-2.664;-2.302)
α^+	23.246	(20.610;26.175)	16.412	(15.024;18.077)
p_1	0.868	(0.830;0.897)		
p_2	0.128	(0.098;0.165)		
p_3	0.005	(0.004;0.005)		
w^*	0.586	(0.541;0.629)		
WAIC	-1291.255		-1081.191	

Table 12.3. *Posterior means and CIs of unknown parameters together with WAIC based on 500 replications for the FDReg and DirReg models. Perturbation factor* $(0.1, 0.8, 0.1)'$

Scenario (iv)	FDReg		DirReg	
	Mean	CI	Mean	CI
$\beta_{10} = 1$	0.735	(0.693;0.778)	0.732	(0.688;0.779)
$\beta_{20} = 0.5$	0.230	(0.151;0.316)	0.279	(0.229;0.326)
$\beta_{11} = 2$	1.856	(1.487;2.328)	1.912	(1.752;2.068)
$\beta_{21} = -3$	-2.934	(-3.425;-2.307)	-2.822	(-2.980;-2.666)
α^+	21.476	(17.476;25.723)	16.953	(15.099;18.750)
p_1	0.884	(0.877;0.893)		
p_2	0.004	(0.003;0.005)		
p_3	0.116	(0.106;0.130)		
w^*	0.577	(0.522;0.634)		
WAIC	-1221.157		-1066.353	

Table 12.4. *Posterior means and CIs of unknown parameters together with WAIC based on 500 replications for the FDReg and DirReg models. Perturbation factor* $(0.1, 0.1, 0.8)'$

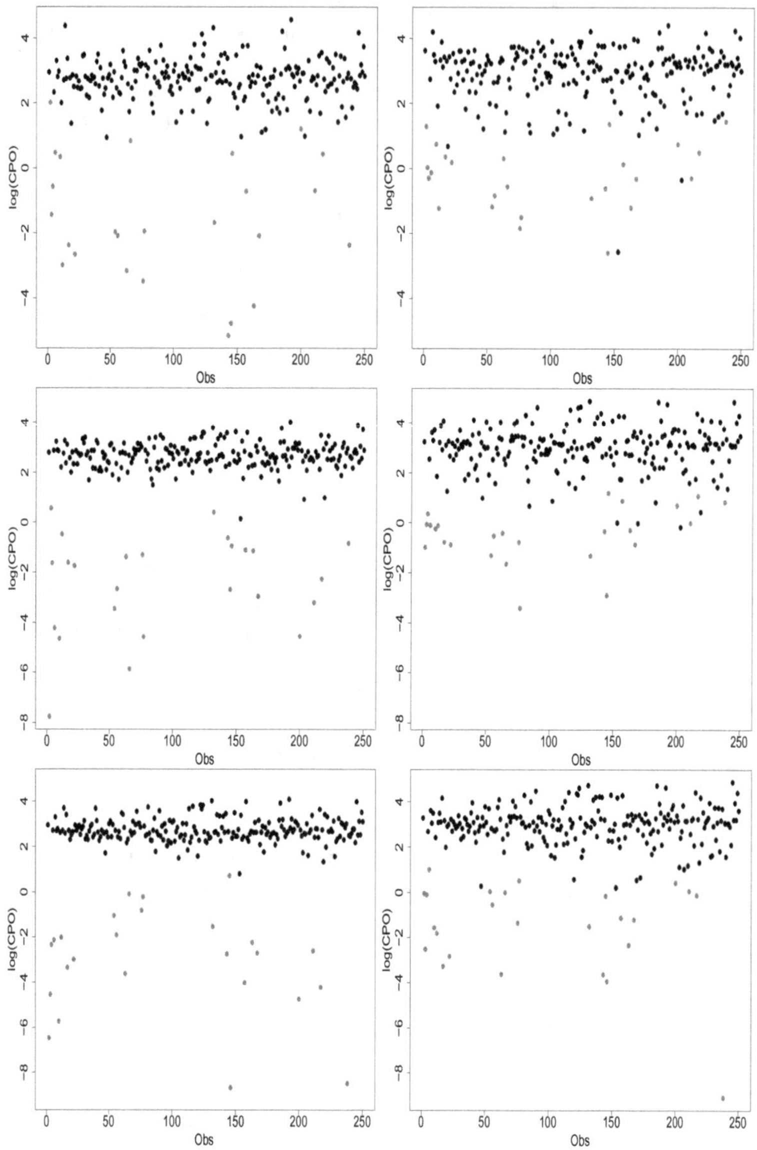

Figure 12.2. *Logarithm of CPO values for the DirReg (left panels) and FDReg (right panels) models in scenarios (ii) (top panel), (iii) (middle panel) and (iv) (bottom panel). CPO values associated with perturbed response values are in red. For a color version of this figure, see www.iste.co.uk/dimotikalis/analysis2.zip*

To evaluate the impact of artificial outliers on the models, we compute the conditional predictive ordinate (CPO) diagnostic based on a cross-validated (leave-one-out) approach (Gelman *et al.* 2014). To avoid the refitting of the model, which would be computationally intensive, we adapt the estimate of CPO (Gelfand and Dey 1994) to the situation at hand: $\widehat{CPO}_i = S\left(\sum_{s=1}^{S} \frac{1}{f(\mathbf{y}_i|\boldsymbol{\eta}^s)}\right)^{-1}$, where S is the number of draws. Please note that $\widehat{CPO}_i$ depends only on the draws $\boldsymbol{\eta}^s$ $(s = 1, \ldots, S)$ from the posterior distributions, and it turns out to be the harmonic mean of the df of $\mathbf{Y}_i$. As a rule of thumb, low values of $\widehat{CPO}_i$ suggest possible influential observations and outliers. Figure 12.2 reports CPO values for both models. Note that scenario (iv) is the most challenging one since the perturbed response values are moved far away from the central body of observations. It follows that the majority of perturbed response values are influential for both models (see the bottom panels of Figure 12.2). Differently, in scenarios (ii) and (iii), the perturbed response values seem to be less influential for the FDReg model than the DirReg model. Thus, the red points, corresponding to CPO values for the perturbed response values, are close to the remaining points in top-right and middle-right panels of Figure 12.2.

12.4.2. *Simulation study 2: generic mixture of two Dirichlet distributions*

We set up a second study with a data generating process following a generic mixture of two Dirichlet distributions sharing a common precision parameter $\alpha^+ = 100$. This simulation mimics the case of a population clustered into two subpopulations with a similar variance but different means, i.e. a case with a latent (unobserved) covariate explaining the clustering. The sample size is $n = 250$ and the mean vector is regressed, according to equation [12.7], onto a quantitative covariate uniformly distributed in $(-0.5, 0.5)$. Regression coefficients are set equal to $\beta_{10} = 1$, $\beta_{11} = 2$, $\beta_{20} = 0.5$, $\beta_{21} = -3$ for the first component of the mixture. Regression coefficients for the second component of the mixture are the same as those of the first component apart from $\beta_{10} = -1$. The mixing proportion of the mixture distribution is equal to 0.7. Please note that the data generating process, a generic mixture of two Dirichlet distributions, does not follow an FD distribution.

The FDReg model provides the best fit to data (see WAIC values in Table 12.5) thanks to the flexibility of the FD distribution to describe bimodal shapes (see the ternary plot in the top-left panel of Figure 12.3). The estimates of regression coefficients are similar across models; nevertheless, the superiority of the FDReg model emerges from the analysis of the behavior of the component means. Figure 12.3 shows, clockwise from top-right panel, the scatterplots of the quantitative covariate x with respect to each element of the composition y_{ij}, $j = 1, 2, 3$. The fitted regression curves of the DirReg (solid lines) and FDReg (dashed lines) models are quite similar. Nevertheless, the FDReg has the component means $\boldsymbol{\lambda}_h$ (dotted curves) as an additional element of flexibility since they are capable of adapting to the clusters induced by the mixture structure of the data generating process. Please note that two dotted curves are represented in each scatterplot since the j–th element, for $j = 1, 2, 3$, of $\boldsymbol{\lambda}_j$ is equal to $\frac{\alpha_j+\tau}{\alpha^++\tau}$, while the j–th elements of the remaining $\boldsymbol{\lambda}_h$, for $h \neq j$, are equal to each other and equal to $\frac{\alpha_j}{\alpha^++\tau}$.

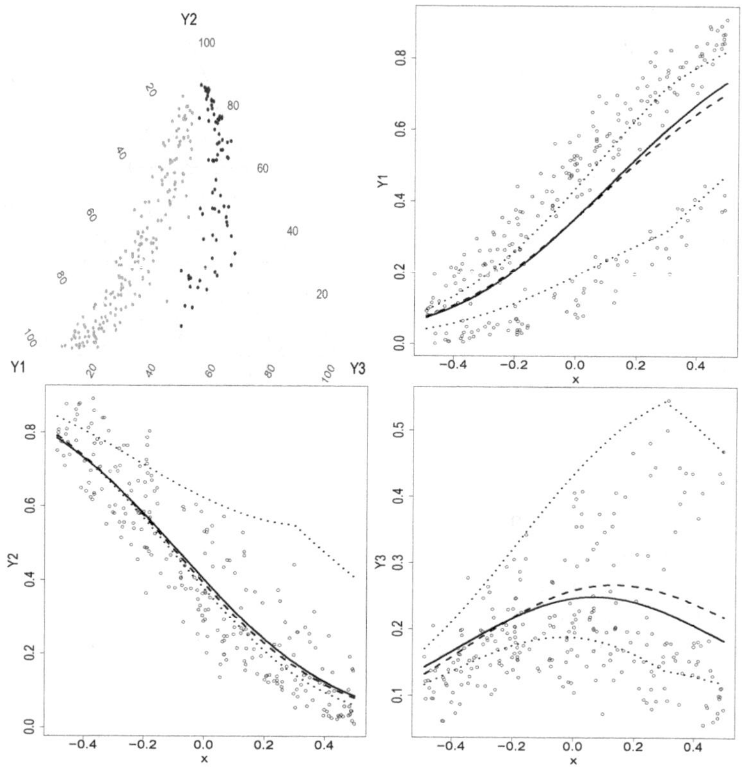

Figure 12.3. *Top-left panel: ternary plot of one simulated sample from simulation study 2. Scatterplots of x versus y_{i1} (top-right panel), y_{i2} (bottom-left panel) and y_{i3} (bottom-right panel). Fitted regression curves for the mean vector of the DirReg model (solid black lines) and the FDReg model (dashed lines). In dotted lines, the regression curves $\boldsymbol{\lambda}_h$ of the FDReg model. For a color version of this figure, see www.iste.co.uk/dimotikalis/analysis2.zip*

12.4.3. *Simulation study 3: FD distribution*

Last, we simulate sample data with $n = 250$ from an FDReg model. The mean vector is regressed onto a quantitative covariate in a similar way to that specified in baseline scenario (i) of simulation study 1. Additional parameters of the FD distribution are $\alpha^+ = 100$, $w^* = 0.6$ and $\mathbf{p} = (1/3, 1/3, 1/3)'$.

The FDReg model ensures the best fit to data and by far the lowest WAIC measure (see Table 12.6). The DirReg model provides acceptable estimates of the regression coefficients since all CIs, apart from the one of β_{21}, contain the true value. Nonetheless, it completely fails to grasp the mixture structure of data as is clear from the graphical representations in Figure 12.4. The $\boldsymbol{\lambda}_h$ component means of the FDReg model perfectly describe the clusters within each element of the composition (dotted lines). Conversely, the mean elements of the DirReg (solid lines) turn out to be almost outside the point clouds. The only element of flexibility available to the DirReg model lies in the modulation of the precision parameter. Indeed, the estimate of the precision parameter α^+ of the DirReg is highly biased downward to induce high variability and allow for describing the separated clusters.

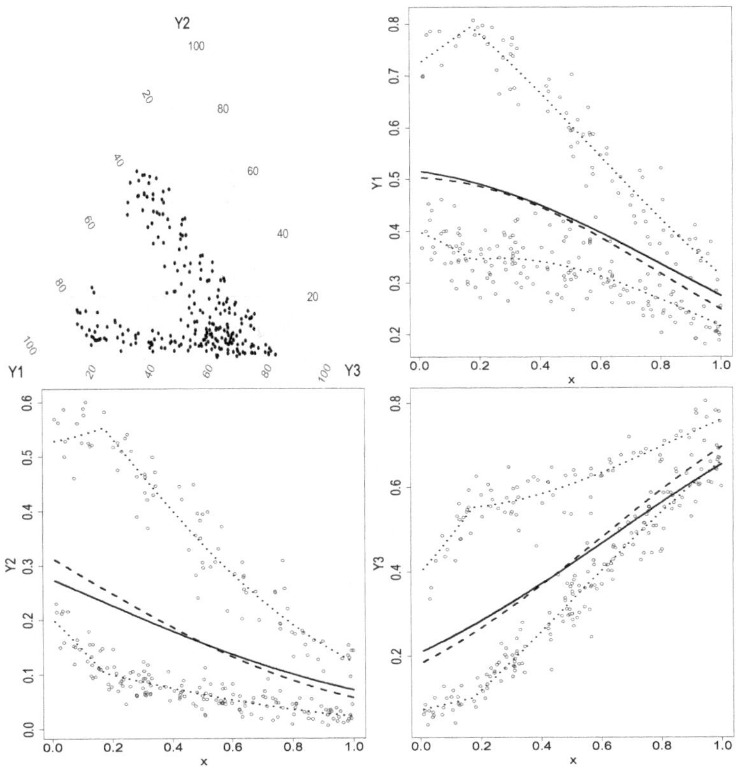

Figure 12.4. *Top-left panel: ternary plot of one simulated sample from simulation study 3. Scatterplots of x versus y_{i1} (top-right panel), y_{i2} (bottom-left panel) and y_{i3} (bottom-right panel). Fitted regression curves for the mean vector of the DirReg model (solid black lines) and the FDReg model (dashed lines). In dotted lines, the regression curves $\boldsymbol{\lambda}_h$ of the FDReg model. For a color version of this figure, see www.iste.co.uk/dimotikalis/analysis2.zip*

	FDReg		DirReg	
	Mean	CI	Mean	CI
β_{10}	0.365	(0.241;0.477)	0.350	(0.311;0.388)
β_{20}	0.509	(0.322;0.679)	0.461	(0.426;0.498)
β_{11}	1.745	(1.524;1.953)	2.008	(1.861;2.147)
β_{21}	-2.540	(-3.119;-2.008)	-2.559	(-2.702;-2.438)
α^+	21.780	(19.916;24.311)	11.569	(10.798;12.385)
p_1	0.628	(0.589;0.673)		
p_2	0.203	(0.023;0.387)		
p_3	0.169	(0.009;0.340)		
w^*	0.506	(0.455;0.558)		
WAIC	-1067.373		-893.853	

Table 12.5. *Simulation study 2: posterior means and CIs of unknown parameters together with WAIC based on 500 replications for the FDReg and DirReg models*

	FDReg		DirReg	
	Mean	CI	Mean	CI
$\beta_{10} = 1$	0.973	(0.840;1.103)	0.998	(0.821;1.177)
$\beta_{20} = 0.5$	0.480	(0.388;0.581)	0.329	(0.092;0.543)
$\beta_{11} = 2$	-1.964	(-2.104;-1.817)	-1.912	(-2.165;-1.676)
$\beta_{21} = -3$	-2.961	(-3.106;-2.823)	-2.622	(-2.966;-2.275)
$\alpha^+ = 100$	91.653	(83.674;96.691)	9.425	(8.996;9.876)
$p_1 = 1/3$	0.324	(0.259;0.385)		
$p_2 = 1/3$	0.331	(0.299;0.367)		
$p_3 = 1/3$	0.345	(0.304;0.382)		
$w^* = 0.6$	0.601	(0.574;0.626)		
WAIC	-1489.116		-818.006	

Table 12.6. *Simulation study 3: posterior means and CIs of unknown parameters together with WAIC based on 500 replications for the FDReg and DirReg models*

12.5. Discussion

The FDReg proves to be a flexible model for compositional data. In addition to its good theoretical properties, we show its adaptability to several scenarios. If data come from a simpler model, such as the DirReg, it provides adequate fit without the risk of over-fitting. Conversely, if data have a clear bimodal structure, the DirReg performs poorly, while the FDReg greatly adapts thanks to its mixture structure. Although not designed as a model to cope with outliers, the FDReg is proven to adapt to a variety of perturbation schemes that induce artificial outlying observations. It provides a better fit and a lower "sensibility" (meaning that perturbed observations are less influential) than the DirReg model. Moreover, the FDReg is computationally very tractable and it has runtimes similar to the DirReg ones despite its greater complexity. It follows that the FDReg should be the preferable model in the presence of a possible bimodality or of influential observations as well as in the absence of a clear mixture structure.

The main limitation of the FDReg lies in the (possibly rigid) assumption that the component means have to be equally far away from each other and with an equal distance proportional to w^*. Possible extensions in this direction will be addressed in future works.

12.6. References

Aitchison, J. (1986). *The Statistical Analysis of Compositional Data.* Chapman and Hall, London.

Campbell, G. and Mosimann, J.E. (1987). Multivariate analysis of size and shape: Modelling with the Dirichlet distribution. *ASA Proceedings of Section on Statistical Graphics*, 93–101.

Di Brisco, A.M. and Migliorati, S. (2017). A special Dirichlet mixture model for multivariate bounded responses. *Cladag 2017 Book of Short Papers*. Universitas Studiorum, Mantova.

Duane, S., Kennedy, A., Pendleton, B.J., and Roweth, D. (1987). Hybrid Monte Carlo. *Physics Letters B*, 195(2), 216–222.

Filzmoser, P. and Hron, K. (2008). Outlier detection for compositional data using robust methods. *Mathematical Geosciences*, 40(3), 233–248.

Frühwirth-Schnatter, S. (2006). *Finite Mixture and Markov Switching Models*. Springer Science & Business Media, New York.

Gelfand, A.E. and Dey, D.K. (1994). Bayesian model choice: Asymptotics and exact calculations. *Journal of the Royal Statistical Society: Series B (Methodological)*, 56(3), 501–514.

Gelman, A., Carlin, J.B., Stern, H.S., and Rubin, D.B. (2014). *Bayesian Data Analysis 2*. Taylor & Francis, New York.

Gueorguieva, R., Rosenheck, R., and Zelterman, D. (2008). Dirichlet component regression and its applications to psychiatric data. *Computational Statistics & Data Analysis*, 52(12), 5344–5355.

Hijazi, R.H. (2003). *Analysis of Compositional Data Using Dirichlet Covariate Models*. The American University, Washington, DC, USA.

Maier, M.J. (2014). Dirichletreg: Dirichlet regression for compositional data in r. Report, Department of Statistics and Mathematics, University of Economics and Business, Vienna.

McCullagh, P. and Nelder, J.A. (1989). *Generalized Linear Models 37*. CRC Press, Boca Raton.

Migliorati, S., Ongaro, A., and Monti, G.S. (2017). A structured Dirichlet mixture model for compositional data: Inferential and applicative issues. *Statistics and Computing*, 27(4), 963–983.

Migliorati, S., Di Brisco, A.M., and Ongaro, A. (2018). A new regression model for bounded responses. *Bayesian Analysis*, 13(3), 845–872.

Neal, R.M. (1994). An improved acceptance procedure for the hybrid Monte Carlo algorithm. *Journal of Computational Physics*, 111(1), 194–203.

Ongaro, A. and Migliorati, S. (2013). A generalization of the Dirichlet distribution. *Journal of Multivariate Analysis*, 114, 412–426.

Pawlowsky-Glahn, V., Egozcue, J.J., and Tolosana-Delgado, R. (2015). *Modeling and Analysis of Compositional Data*. John Wiley & Sons, New York.

Stan Development Team (2016). *Stan Modeling Language Users Guide and Reference Manual*. CreateSpace Independent Publishing Platform, Scotts Valley, CA.

Vehtari, A., Gelman, A., and Gabry, J. (2017). Practical Bayesian model evaluation using leave-one-out cross-validation and waic. *Statistics and Computing*, 27(5), 1413–1432.

13

On Asymptotic Structure of the Critical Galton–Watson Branching Processes with Infinite Variance and Allowing Immigration

We observe Galton–Watson branching processes with possible immigration. The main results of this chapter are as follows. In the absence of immigration, an integral form of the generating function of the invariant measure in its domain of definition is obtained. In the existing literature, only the "local" form of this function in the neighborhood of point 1 was known (see Slack (1968)). For the processes with immigration, we establish two theorems. The first establishes a formula showing the asymptotic form of the generating function of transition probabilities. This generalizes the result of Pakes (1975), in the sense that he found a similar formula, but only at point 1. In Theorem 13.3, we find the rate of convergence to invariant measures for processes with an infinite variance of the individual transformation law and an infinite mean of the individual immigration law.

13.1. Introduction

Let $\{X_n, n \in \mathbb{N}_0\}$ be the Galton–Watson branching process allowing immigration (GWPI), where $\mathbb{N}_0 = \{0\} \cup \mathbb{N}$ and $\mathbb{N} = \{1, 2, \ldots\}$. This is a homogeneous discrete-time Markov chain with state space $\mathcal{S} \subset \mathbb{N}_0$ and whose transition probabilities are

$$p_{ij} = \text{coefficient of } s^j \text{ in } h(s)\big(f(s)\big)^i, \quad s \in [0,1),$$

where $h(s) = \sum_{j \in \mathcal{S}} h_j s^j$ and $f(s) = \sum_{j \in \mathcal{S}} p_j s^j$ are probability generating functions (PGFs). The variable X_n is interpreted as the population size in GWPI at the moment n. An evolution of the process will occur by the following scheme. An

Chapter written by Azam A. IMOMOV and Erkin E. TUKHTAEV.

initial state is empty, that is, $X_0 = 0$, and the process starts due to immigrants. Each individual at time n produces j progeny with probability p_j independently of each other so that $p_0 > 0$. Simultaneously, in the population i, immigrants arrive with probability h_i in each moment $n \in \mathbb{N}$. These individuals undergo further transformation, obeying the reproduction law $\{p_j\}$, and n-step transition probabilities $p_{ij}^{(n)} := \mathsf{P}\{X_{n+k} = j \,|\, X_k = i\}$ for any $k \in \mathbb{N}$ are given by

$$\mathcal{P}_n^{(i)}(s) := \sum_{j \in \mathcal{S}} p_{ij}^{(n)} s^j = \left(f_n(s)\right)^i \prod_{k=0}^{n-1} h\left(f_k(s)\right) \quad \text{for any } i \in \mathcal{S}, \tag{13.1}$$

where $f_n(s)$ is the n-fold iteration of PGF $f(s)$; see, for example, Pakes (1979). Note that function $f_n(s)$ generates the distribution law of the number of individuals at the time n in the process without immigration (see section 13.2). Thus, the transition probabilities $\left\{p_{ij}^{(n)}\right\}$ are completely defined by the probabilities $\{p_j\}$ and $\{h_j\}$.

Classification of states of the chain $\{X_n\}$ is one of the fundamental problems in the theory of GWPI. Direct differentiation of [13.1] gives

$$\mathsf{E}\left[X_n | X_0 = i\right] = \begin{cases} an + i & , \quad \text{when } m = 1, \\ \left(\dfrac{a}{m-1} + i\right) m^n - \dfrac{a}{m-1}, & \text{when } m \neq 1, \end{cases}$$

where $m := f'(1-) = \sum_{j \in \mathcal{S}} j p_j$ is the mean per-capita offspring number and $a := h'(1-) = \sum_{j \in \mathcal{S}} j h_j$ is the average number of immigration distribution law. The formula obtained for $\mathsf{E}\left[X_n | X_0 = i\right]$ shows that the classification of states of GWPI depends on a value of the parameter m. Process $\{X_n\}$ is classified as sub-critical, critical and supercritical if $m < 1$, $m = 1$ and $m > 1$, respectively.

The population process described above was first considered by Heathcote in 1965 (Heatcote 1965). Further long-term properties of $\mathcal{S}$ and a problem of existence and uniqueness of invariant measures of GWPI were investigated by Seneta (1969), Pakes (1971a, 1971b) and by many other authors. Therein, some moment conditions for PGF $f(s)$ and $h(s)$ were required to be satisfied. In Seneta's aforementioned works, the ergodic properties of $\{X_n\}$ were investigated. He has proved that when $m \leq 1$, the process $\{X_n\}$ has an invariant measure $\{\mu_k,\, k \in \mathcal{S}\}$, which is unique up to multiplicative constant. Pakes (1971b) has shown that in the supercritical case, $\mathcal{S}$ is transient. In the critical case, $\mathcal{S}$ can be transient, null-recurrent or ergodic. In this case, if we assume, in addition, that $2b := f''(1-) < \infty$, properties of $\mathcal{S}$ depend on the value of parameter $\lambda = a/b$: if $\lambda > 1$ or $\lambda < 1$, then $\mathcal{S}$ is transient or null-recurrent accordingly. In the case when $\lambda = 1$, Pakes (1971a) studied necessary and sufficient conditions for a null-recurrence property. Limiting distribution law for critical process $\{X_n\}$ was found first by Seneta (1970). He has proved that the normalized process $X_n/(bn)$ has limiting Gamma distribution with density function

$\Gamma^{-1}(\lambda)x^{\lambda-1}e^{-x}$, provided that $0 < \lambda < \infty$, where $x > 0$ and $\Gamma(*)$ is Euler's Gamma function. This result has been established also by Pakes (1971a) without reference to Seneta. Afterwards, Pakes (1979, 1975), has obtained principally new results for all cases $m < \infty$ and $b = \infty$.

Throughout this chapter, we keep on the critical case only and $b = \infty$. Our reasoning will be bound up with elements of slow variation theory in the sense of Karamata; see Seneta (1972). We remember that the real-valued, positive and measurable function $L(x)$ is said to be slowly varying (SV) at infinity if $L(\lambda x)/L(x) \to 1$ as $x \to \infty$ for each $\lambda > 0$. For more information, see Seneta (1972), Asmussen (1983) and Bingham (1987).

In section 13.2, we study invariant measures of the simple Galton–Watson (GW) process $\{Z_n\}$. In Theorem 13.1 an integral form of PGF $U(s)$ of an invariant measure of the process $\{Z_n\}$ and an asymptotic form of derivative $U'(s)$ neighborhood of point 1 appear. This theorem expands Slack's result (1968), in the sense that he found only local representation of the function $U(s)$ in a neighborhood of point 1.

In section 13.3, we investigate invariant properties of GWPI. We observe an asymptotic expansion of $\mathcal{P}_n(s) := \mathcal{P}_n^{(0)}(s)$, supposing that $h'(1-) = \sum_{j\in\mathcal{S}} jh_j = \infty$ but PGF $h(s)$ regularly varies (see the representation $[h_\delta]$ below).

13.2. Invariant measures of GW process

Let $\{Z_n, n \in \mathbb{N}_0\}$ be the simple GW branching process without immigration, given by offspring PGF $f(s)$. Discussing this case, we will assume that the offspring PGF $f(s)$ has the following representation:

$$f(s) = s + (1-s)^{1+\nu}\mathcal{L}\left(\frac{1}{1-s}\right),] \qquad [f_\nu]$$

where $0 < \nu < 1$ and $\mathcal{L}(x)$ is SV at infinity. By the criticality of the process, the condition $[f_\nu]$ implies that $b = \infty$. This includes the case $b < \infty$ when $\nu = 1$ and $\mathcal{L}(t) \to b$ as $t \to \infty$.

Consider PGF $f_n(s) := \mathsf{E}\left[s^{Z_n} | Z_0 = 1\right]$ and write $R_n(s) := 1 - f_n(s)$. Evidently, $Q_n := R_n(0)$ is the survival probability of the process. By using Slack's arguments (1968), we can show that if the condition $[f_\nu]$ holds, then

$$Q_n^\nu \cdot \mathcal{L}\left(\frac{1}{Q_n}\right) \sim \frac{1}{\nu n} \quad asn \to \infty. \qquad [13.2]$$

Slack (1968) also has shown that

$$\mathcal{U}_n(s) := \frac{f_n(s) - f_n(0)}{f_n(0) - f_{n-1}(0)} \longrightarrow U(s) \qquad [13.3]$$

for $s \in [0,1)$, where the limit function $U(s)$ satisfies the Abel equation

$$U(f(s)) = U(s) + 1, \qquad [13.4]$$

so that $U(s)$ is the PGF of invariant measure for the GW process $\{Z_n\}$. Combining $[f_\nu]$, [13.2] and [13.3] and considering the properties of the process $\{Z_n\}$, we have

$$\mathcal{U}_n(s) \sim U_n(s) := \left[1 - \frac{R_n(s)}{Q_n}\right] \nu n \quad as\ n \to \infty.$$

So we proved the following lemma.

LEMMA 13.1.– *If the condition $[f_\nu]$ holds, then*

$$R_n(s) = \frac{\mathcal{N}(n)}{(\nu n)^{1/\nu}} \cdot \left[1 - \frac{U_n(s)}{\nu n}\right], \qquad [13.5]$$

where the function $\mathcal{N}(x)$ is SV at infinity and

$$\mathcal{N}(n) \cdot \mathcal{L}^{1/\nu}\left(\frac{(\nu n)^{1/\nu}}{\mathcal{N}(n)}\right) \longrightarrow 1 \quad as\ n \to \infty, \qquad [13.6]$$

and the function $U_n(s)$ satisfies the following properties:

– $U_n(s) \longrightarrow U(s)$ *as $n \to \infty$ so that the equation* [13.4] *holds;*

– $\lim_{s\uparrow 1} U_n(s) = \nu n$ *for each fixed $n \in \mathbb{N}$;*

– $U_n(0) = 0$ *for each fixed $n \in \mathbb{N}$.*

Apparently, this lemma is the generalization of [13.2], which is established with a simpler proof than that shown in Imomov (2019).

Furthermore:

$$\Lambda(y) := \frac{f(1-y) - (1-y)}{y} = y^\nu \mathcal{L}\left(\frac{1}{y}\right),$$

we establish the following important assertion.

LEMMA 13.2.– *If the condition $[f_\nu]$ holds, then*

– *the following relation is true:*

$$\frac{\partial R_n(s)}{\partial s} = -\psi_n(s) \frac{R_n(s)\Lambda(R_n(s))}{(1-s)\Lambda(1-s)},$$

where $\psi_n(s)$ is continuous increasing on $s \in [0,1]$, for all $n \in \mathbb{N}$ and

$$\frac{f'(s)}{f'(f_n(s))} < \psi_n(s) < 1;$$

– *the following asymptotic relation is true:*

$$\frac{\partial R_n(s)}{\partial s} \sim -\psi(s)\frac{R_n(s)\Lambda\big(R_n(s)\big)}{(1-s)\Lambda(1-s)} \quad as \;\; n \to \infty,$$

where $\psi(s)$ *is continuous increasing on* $s \in [0,1]$*, so that*

$$f'(s) \le \psi(s) \le 1;$$

– *the following locally asymptotic relation is true:*

$$\frac{\partial R_n(s)}{\partial s} = \frac{R_n(s)\Lambda\big(R_n(s)\big)}{(1-s)\Lambda(1-s)}\big(1+\phi(1-s)\big) \quad as \;\; s \uparrow 1 \;\; and \;\; n \to \infty,$$

where $\phi(y) = -(1+\nu)\Lambda(y)\big(1+o(1)\big)$ *as* $y \downarrow 0$.

The statements of the last lemma will play an important role in the proof of Theorem 13.1.

Now consider the function

$$\mathcal{M}_n(s) = 1 - \frac{\Lambda\big(R_n(s)\big)}{\Lambda\big(Q_n\big)}. \qquad [13.7]$$

It follows from [13.5] and from SV properties that

$$\mathcal{M}_n(s) = 1 - \left(\frac{R_n(s)}{Q_n}\right)^{\nu} \frac{\mathcal{L}\big(1/R_n(s)\big)}{\mathcal{L}\big(1/Q_n\big)}$$

$$\sim 1 - \left(1 - \frac{U_n(s)}{\nu n}\right)^{\nu} = \frac{U_n(s)}{n}\big(1+\kappa_n(s)\big) \quad as \;\; n \to \infty,$$

where $\kappa_n(s) = \mathcal{O}\big(1/n\big)$ uniformly in $s \in [0,1)$.

Thus, we have the following:

LEMMA 13.3.– *If the condition* $[f_\nu]$ *holds, then*

$$n \cdot \mathcal{M}_n(s) \longrightarrow U(s) \quad as \;\; n \to \infty, \qquad [13.8]$$

where $U(s)$ *is the PGF of invariant measure of GW process.*

The following statement gives an asymptotical representation for $\Lambda\big(R_n(s)\big)$ and can be substituted for Lemma 13.1.

LEMMA 13.4.– *If the condition $[f_\nu]$ holds, then*

$$\Lambda(R_n(s)) = \frac{\Lambda(1-s)}{\Lambda(1-s)\,\nu n+1}\left(1+o(1)\right) \quad as\ n\to\infty. \qquad [13.9]$$

REMARK 13.1.– *The asymptotic relation* [13.9] *seems, in appearance, to be an analog of classical form of the basic lemma of the theory of critical GW branching processes with finite variance, in which $b = f''(1-)/2$ instead of ν and $\Lambda(x)\equiv x$; see, for instance, Imomov (2020, Lemma 1 (ii)).*

REMARK 13.2.– *Along with all applications, the second statement of Lemma 13.2 in combining with the formula* [13.9] *provides an opportunity to find an asymptotic representation of the transition probability $P_{11}(n) := \mathsf{P}\{Z_n = 1 \,|\, Z_0 = 1\}$ as $n\to\infty$, since $f_n'(0) = P_{11}(n)$. In fact, we obtain*

$$P_{11}(n) \sim \frac{\psi(0)}{p_0}\cdot\frac{\mathcal{N}(n)}{(\nu n)^{1+1/\nu}} \quad as\ n\to\infty,$$

where $p_1 < \psi(0) < 1$ and $\mathcal{N}(\cdot)$ is SV, defined in [13.6].

Now using [13.5]–[13.9] and considering Lemma 13.2, we track down an explicit form of PGF $U(s)$ and the asymptote of its derivative.

THEOREM 13.1.– *If the condition $[f_\nu]$ holds, then*

– *the PGF $U(s)$ is the form of*

$$U(s) = \int_0^s \frac{\psi(u)}{(1-u)\Lambda(1-u)}\,du,$$

where $\psi(s)$ is continuous increasing on $s\in[0,1]$, so that

$$f'(s) \le \psi(s) \le 1;$$

– *the function $U'(s)$ has the following locally asymptotic form:*

$$U'(s) = \frac{1}{(1-s)\Lambda(1-s)}\left(1+\phi(1-s)\right) \quad as\ s\uparrow 1,$$

where $\phi(y) = -(1+\nu)\Lambda(y)\left(1+o(1)\right)$ as $y\downarrow 0$.

13.3. Invariant measures of GWPI

In this section, we consider GWPI. First of all, we recall the following theorem, which was proved by Pakes (1975).

THEOREM P1 (Pakes 1975).– *If* $m = 1$, *then*

$$p_{00}^{(n)} \sim K \exp\left\{ \int_1^{e^n} \frac{\ln h(1-\varphi(y))}{y} dy \right\} \quad as\ n \to \infty,$$

where $\varphi(y)$ *is the decreasing SV-function. If*

$$\sum_{m=0}^{\infty} \Big[(1 - h(f_m(0))\, (1 - f'(f_m(0)) \Big] < \infty,$$

then

$$p_{00}^{(n)} \sim K_1 \exp\left\{ \int_0^{f_n(0)} \frac{\ln h(y)}{f(y) - y} dy \right\} \quad as\ n \to \infty.$$

Herein, K *and* K_1 *are some constants.*

Since this point, everywhere we will consider the case that immigration PGF $h(s)$ has the following form:

$$1 - h(s) = (1-s)^{\delta} \ell\left(\frac{1}{1-s}\right),] \qquad [h_\delta]$$

where $0 < \delta < 1$ and $\ell(x)$ is SV at infinity. The assumption $[h_\delta]$ implies that an average number of immigration distribution law is infinite, i.e. $\sum_{j \in \mathcal{S}} j h_j = \infty$, but $\sum_{j \in \mathcal{S}} j^{\delta} h_j < \infty$.

Our results appear provided that conditions $[f_\nu]$ and $[h_\delta]$ hold and $\delta > \nu$. As has been shown in Pakes (1975), in this case, $\mathcal{S}$ is ergodic. Namely, we improve statements of Theorem P1. Here, we put forward an additional requirement concerning $\mathcal{L}(x)$ and $\ell(x)$. Since $\mathcal{L}(x)$ is SV, we can write

$$\frac{\mathcal{L}(\lambda x)}{\mathcal{L}(x)} = 1 + \alpha(x)] \qquad [\mathcal{L}_\alpha]$$

for each $\lambda > 0$, where $\alpha(x) \to 0$ as $x \to \infty$. Henceforth, we suppose that some positive function $g(x)$ is given so that $g(x) \to 0$ and $\alpha(x) = o\big(g(x)\big)$ as $x \to \infty$. In this case, $\mathcal{L}(x)$ is called SV with remainder $\alpha(x)$; see Bingham (1987, p. 185, condition SR3). Wherever we exploit the condition $[\mathcal{L}_\alpha]$, we will suppose that

$$\alpha(x) = o\left(\frac{\mathcal{L}(x)}{x^{\nu}}\right) \quad as\ x \to \infty. \qquad [13.10]$$

Moreover, by perforce, we suppose the condition

$$\frac{\ell(\lambda x)}{\ell(x)} = 1 + \beta(x)] \qquad [\ell_\beta]$$

for each $\lambda > 0$, where

$$\beta(x) = o\left(\frac{\ell(x)}{x^{\delta}}\right) \quad as \ x \to \infty.$$

Since $f_n(s) \uparrow 1$ for all $s \in [0,1)$ in virtue of [13.1], it is sufficient to observe the case $i = 0$ as $n \to \infty$. Denote

$$\mathcal{P}_n(s) := \mathcal{P}_n^{(0)}(s).$$

The following theorem is a generalization of Theorem P1.

THEOREM 13.2.– *Let conditions* $[f_\nu]$, $[h_\delta]$ *hold. If* $\delta > \nu$, *then*

$$\mathcal{P}_n(s) \sim K(s)\exp\left\{-\int_s^{f_n(s)} \frac{1-h(y)}{f(y)-y}\Big[1+\delta(1-y)\Big]dy\right\}$$

as $n \to \infty$, *where* $K(s)$ *is a bounded function for* $s \in [0,1)$ *and* $\delta(x) \to 0$ *as* $x \downarrow 0$. *If in addition, the conditions* $[\mathcal{L}_\alpha]$ *and* [13.10] *are satisfied, then*

$$\delta(x) = \mathcal{O}\big(\Lambda(x)\big) \quad as \ x \downarrow 0.$$

The next result directly follows from Theorem 13.2 setting $x = 0$ there.

COROLLARY 13.1.– *Let conditions* $[f_\nu]$, $[h_\delta]$ *hold. If* $\delta > \nu$, *then*

$$p_{00}^{(n)} \sim A\exp\left\{-\frac{1}{\delta-\nu}\cdot L\left(\frac{(\nu n)^{1/\nu}}{\mathcal{N}(n)}\right)\right\} \quad as \ n \to \infty,$$

where A *is a positive constant,* $L := \ell/\mathcal{L}$ *and* $\mathcal{N}(x)$ *is SV defined in* [13.6].

Further, we need the following result which is an improved analog of the basic lemma of the theory of critical GW processes.

LEMMA 13.5 (Imomov and Tukhtaev 2019).– *Let conditions* $[f_\nu]$, $[\mathcal{L}_\alpha]$ *and* [13.10] *hold. Then*

$$\frac{1}{\Lambda\big(R_n(s)\big)} - \frac{1}{\Lambda(1-s)} = \nu n + \frac{1+\nu}{2}\cdot\ln\big(1+\nu n\Lambda(1-s)\big) + \rho_n(s),$$

where $\rho_n(s) = o(\ln n) + \sigma_n(s)$ *and* $\sigma_n(s)$ *is bounded uniformly in* $s \in [0,1)$ *and converges to the limit* $\sigma(s)$ *as* $n \to \infty$, *which is a bounded function for* $s \in [0,1)$.

We make sure that at the conditions of the second part of Theorem 13.2, PGF $\mathcal{P}_n(s)$ converges to a limit $\pi(s)$, which we denote by the power series representation

$$\pi(s) = \sum_{j\in\mathcal{S}} \pi_j s^j.$$

Now, using Lemma 13.5, we can establish a speed rate of this convergence in the following theorem.

THEOREM 13.3.– *Let conditions* $[f_\nu]$, $[h_\delta]$ *hold and* $\delta > \nu$. *Then,* $\mathcal{P}_n(s)$ *converges to* $\pi(s)$ *which generates the invariant measures* $\{\pi_j\}$ *for GWPI. The convergence is uniform over compact subsets of the open unit disc. If, in addition, the conditions* $[\mathcal{L}_\alpha]$, [13.10] *and* $[\ell_\beta]$ *are fulfilled, then*

$$\mathcal{P}_n(s) = \pi(s)\left(1+\Delta_n(s)\mathcal{N}_\delta\left(\frac{1}{R_n(s)}\right)\right),$$

where $\mathcal{N}_\delta(x) = \mathcal{N}^\delta(x)\ell(x)$, *the function* $\mathcal{N}(x)$ *is defined in* [13.6], *herein*

$$\Delta_n(s) = \frac{1}{\delta-\nu}\frac{1}{\left(\nu_n(s)\right)^{\delta/\nu-1}} - \frac{1+\nu}{2\nu}\frac{\ln\left[\nu_n(s)\right]}{\left(\nu_n(s)\right)^{\delta/\nu}}(1+o(1))$$

as $n\to\infty$ *and* $\nu_n(s) = \nu n + \Lambda^{-1}(1-s)$.

The following result is the direct consequence of Theorem 13.3.

COROLLARY 13.2.– *If conditions of Theorem 13.3 hold, then*

$$p_{00}^{(n)} = \pi_0\cdot\left(1+\Delta_n\mathcal{N}_\delta(n)\right),$$

where $\mathcal{N}_\delta(n)$ *is SV at infinity and*

$$\Delta_n = \frac{1}{\delta-\nu}\frac{1}{(\nu n)^{\delta/\nu-1}} - \frac{1+\nu}{2\nu}\frac{\ln n}{(\nu n)^{\delta/\nu}}(1+o(1)) \quad \text{as } n\to\infty.$$

REMARK 13.3.– *The analogous result, as in Theorem 13.2, has been proved in Imomov (2015) for* $\delta = 1$ *and* $f'''(1-) < \infty$.

13.4. Conclusion

In this chapter, we consider and study the model of the evolution of the population size of homogeneous individuals, called the branching process allowing immigration. The main goal of the work is to study the asymptotic properties of the process trajectory in the interpretation of transition probabilities in the minimal moment conditions.

In the monograph (Harris 1963, pp. 29–31), part of the treatment of gene fixation was interpreted in terms of invariant (stationary) measures. We hope that the properties of the invariant measures of the simple GW branching process established in Theorem 13.1, and the asymptotic formulas for transition probabilities for the GWPI (Theorem 13.2 and Theorem 13.3), showing approximations to invariant measures, can be useful in theoretical aspects of applied problems similar to those described in Harris (1963).

13.5. References

Asmussen, S. and Hering, H. (1983). *Branching Processes*. Birkhäuser, Boston.

Bingham, N.H., Goldie, C.M., Teugels, J.L. (1987). *Regular Variation*. Cambridge University Press, Cambridge.

Harris, T.E. (1963). *The Theory of Branching Processes*. Springer-Verlag, Berlin.

Heathcote, C.R. (1965). A branching process allowing immigration. *Jour. Royal Stat. Soc.*, B-27, 138–143.

Imomov, A.A. (2015). On long-time behaviors of states of Galton–Watson branching processes allowing immigration. *Jour. Siber. Fed. Univ. Math. Phys.*, 8(4), 394–405.

Imomov, A.A. (2019). On a limit structure of the Galton–Watson branching processes with regularly varying generating functions. *Prob. and Math. Stat.*, 39(1), 61–73.

Imomov, A.A. (2020). Renewed limit theorems for the discrete-time branching process and its conditioned limiting law interpretation [Online]. Available at: https://arxiv.org/abs/2004.09307.

Imomov, A.A. and Tukhtaev, E.E. (2019). On application of slowly varying functions with remainder in the theory of Galton–Watson branching process. *Jour. Siber. Fed. Univ. Math. Phys.*, 12(1), 51–57.

Pakes, A.G. (1971a). On the critical Galton–Watson process with immigration. *Jour. Austral. Math. Soc.*, 12, 476–482.

Pakes, A.G. (1971b). Branching processes with immigration. *Jour. Appl. Prob.*, 8(1), 32–42.

Pakes, A.G. (1975). Some results for non-supercritical Galton–Watson processes with immigration. *Math. Biosci.*, 24, 71–92.

Pakes, A.G. (1979). Limit theorems for the simple branching processes allowing immigration I: The case of finite offspring mean. *Adv. Appl. Prob.*, 11, 31–62.

Seneta, E. (1969). Functional equations and the Galton–Watson process. *Adv. Appl. Prob.*, 1, 1–42.

Seneta, E. (1970). An explicit-limit theorem for the critical Galton–Watson process with immigration. *Jour. Royal Stat. Soc.*, B-32(1), 149–152.

Seneta, E. (1972). *Regularly Varying Functions*. Springer, Berlin.

Slack, R.S. (1968). A branching process with mean one and possible infinite variance. *Wahrscheinlichkeitstheor. und Verv. Geb.*, 9, 139–145.

14

Properties of the Extreme Points of the Joint Eigenvalue Probability Density Function of the Wishart Matrix

We will examine some properties of the extreme points of the probability density distribution of the Wishart matrix, using properties of the Vandermonde determinant and showing examples of the applications of these properties.

14.1. Introduction

The Gaussian ensembles of random matrices have been extensively investigated and date back to the works on the statistical distribution of widths and spacings of nuclear resonance levels (Wigner 1951) and the statistical theory of the energy levels of complex systems (Dyson 1962). Thus, random matrix theory (as thoroughly discussed in Mehta (1967), König (2005) and Forester (2010)) has proved to be pivotal in high dimensional and/or multivariate statistical analysis, plus many other applications based on the Wishart matrix (Anderson 2003), as well as orthogonal polynomials (Szegő 1939). Therefore, we attempt to investigate the properties of the extreme points of the joint eigenvalue probability density function of the random Wishart matrix optimized over the unit p-sphere (Muhumuza *et al.* 2018b). We also apply the techniques of Vandermonde polynomial optimizations (Lundengård and Silvestrov 2013), matrix factorization (Oruç and Phillips 2000) and eigenvalue optimization (Golub and Van Loan 1996), matrix norms (Demmel 1997) and condition number (Edelman 1988).

Chapter written by Asaph Keikara MUHUMUZA, Karl LUNDENGÅRD, Sergei SILVESTROV, John Magero MANGO and Godwin KAKUBA.

Investigating the characteristic properties of the extreme or optimal points of the joint eigenvalue probability density functions proves to be of great significance, especially in optimal regression, polynomial interpolation and numerical approximation (Phillips 2003), stability and sensitivity analysis in numerical computing (Hoel 1958), optimal experimental design (Karlin and Studden 1966) and optimal control theory in stochastic and random processes (Ljung *et al.* 2012).

The aim of this chapter is to illustrate the significance of the extreme points of the Vandermonde matrix, in order to optimize its condition number as a measure of sensitivity and stability of the given system. This will be discussed in section 14.5, but first we give a brief outline on the background of the problem setup, based on polynomial regression models and the close relation between the Vandermonde matrix and the random Wishart matrix.

14.2. Background

For convenience, we will use the following notation for the Vandermonde matrix of size $N \times N$:

$$\mathbf{X} = \mathbf{V}_N(\mathbf{x}) = \begin{bmatrix} 1 & 1 & \cdots & 1 \\ x_1 & x_2 & \cdots & x_N \\ \vdots & \vdots & \ddots & \vdots \\ x_1^{N-1} & x_2^{N-1} & \cdots & x_N^{N-1} \end{bmatrix}.$$

Here, $\mathbf{x} = (x_1, x_2, \ldots, x_N)$ are N distinct data points or nodes. Note that the Vandermonde matrix has a simple expression for its determinant given by

$$v_n(\mathbf{x}) \equiv |\mathbf{X}| = \prod_{1 \leq i < j \leq N} (x_i - x_j) \qquad [14.1]$$

where $|\cdot|$ denotes the determinant.

The usefulness of the Vandermonde matrix cannot be overemphasized, especially in polynomial interpolation, approximation and nonlinear regression models that can be expressed in the form $\mathbf{Y} = \mathbf{X}\boldsymbol{\beta} + \boldsymbol{\varepsilon}$, where $\mathbf{X}$ is the Vandermonde matrix, $\mathbf{Y} = (y_1, y_2, \ldots, y_N)^\top$ is the response vector and the parameter vector, $\boldsymbol{\beta} = (\beta_1, \beta_2, \ldots, \beta_N)^\top$ and the vector $\boldsymbol{\varepsilon} = (\varepsilon_1, \varepsilon_2, \ldots, \varepsilon_N)^\top$ are random errors for which $\varepsilon_i \sim \mathcal{N}(0, \sigma^2)$. Here, the symbol $^\top$ represents the transpose (Phillips 2003).

Since the entries of the Vandermonde matrix are monomials of the form $x_i^{j-1}, i, j = 1(1)N$, then the Wishart matrix or moment matrix $\mathbf{W} = \mathbf{X}^\top\mathbf{X}$

has entries that are also polynomials of the form $x_{ij} = \sum_{i,j=1}^{N} x_i^{j-1}$ which can also be taken as entries of a Hankel matrix (Ljung *et al.* 2012). Applying the usual Newton–Girard symmetric function (Abramowitz and Stegun 1965; Macdonald 1979), the matrices $\mathbf{X}$ and $\mathbf{W}$ can be decomposed and directly evaluate their inverses, eigenvalues, determinants, matrix norms and condition numbers, as well as explain some characteristic properties of extreme points of their determinants and other applications. By extreme points, we refer to those points of the Vandermonde matrix that maximize its determinant (as fully discussed in Muhumuza *et al.* (2018a); Lundengård and Silvestrov (2013)). Knowing the extreme values of the Vandermonde determinant can assist us in estimating the conditional number of the Vandermonde matrix and the Wishart matrix. In sections 14.3 and 14.4, we describe how some properties of the Vandermonde matrix and Wishart matrix relate to one another and in section 14.5 we apply these relations when computing the conditional number of the two types of matrices.

14.3. Polynomial factorization of the Vandermonde and Wishart matrices

The principles of the Vandermonde matrix factorization using both elementary symmetric and complete symmetric functions have previously been studied, especially by Bjorck and Pereyra (1970), Bjorck and Elfving (1973), Tang and Golub (1981), Golub and Van Loan (1996), Martinez and Pena (1998), El-Mikkawy (2003), El-Mikkawy and El-Desouky (2003), Yang and Qiao (2003), Oruç and Akmaz (2004), Yang (2005, 2007a, b), Oruç and Phillips (2007) and Spivey and Zimmer (2008).

DEFINITION 14.1.– *The Newton–Girard elementary symmetric functions $e_k(x_0, x_1, \ldots, x_N)$ are given by*

$$e_\tau(x_0, x_1, \ldots, x_N) = \sum_{0 \le k_1 < k_2 < \ldots < k_\tau \le N} x_{k_1} x_{k_2} \cdots x_{k_\tau} \qquad [14.2]$$

and the complete symmetric functions $h_\tau(x_0, x_1, \ldots, x_N)$ are given by

$$h_\tau(x_0, x_1, \ldots, x_N) = \sum_{0 \le k_1 < k_2 < \ldots < k_\tau \le N} x_{k_1}^{r_1} x_{k_2}^{r_2} \cdots x_{k_\tau}^{r_N}, \qquad [14.3]$$

where $r_1 + r_2 + \ldots + r_N = \tau$, $r_1, r_2, \ldots, r_N \in \{0, 1, \ldots, \tau\}$, $e_\tau(\mathbf{x}) = h_\tau(\mathbf{x}) = 0$ for $r < 0$ and $e_\tau(\mathbf{x}) = h_\tau(\mathbf{x}) = 1$ for $r = 0$. These symmetric functions have generation functions given by

$$E(t) = \sum_{\tau=0}^{N+1} e_\tau(x_0, x_1, \ldots, x_N) t^\tau = \prod_{i=0}^{N} (1 - x_i t)$$

and

$$H(t) = \frac{1}{E(t)} = \sum_{\tau=0}^{N+1} h_\tau(x_0, x_1, \ldots, x_N)t^\tau = \prod_{i=0}^{N}(1 - x_i t)^{-1}$$

where $H(t)E(t) = 1$ *and* $\sum_{r=0}^{m}(-1)^{m-\tau} e_{m-\tau}(x)h_{N-l}(x) = 0,\ n \geq l.$

Let $\mathbf{X} = \mathbf{V}_N(\mathbf{x}) = \{x_j^i\}_{i,j=1}^{N}$, be an $N \times N$ where $\mathbf{x} = (x_1, x_2, \ldots, x_N)$ are pairwise distinct points. Setting $\mathbf{P}_N[x]$ to be a vector space of polynomials in x over the field $\mathbb{R}$ of degree, at most, N, then we define the sets $\mathbb{B}_1 = \{1, x, x^2, \ldots, x^N\}$ and $\mathbb{B}_2 = \{[x]_0, [x]_1, \ldots, [x]_N\}$; both form the bases of $\mathbf{P}_N[x]$, where $[x]_k = (x - x_0)(x - x_1)\ldots(x - x_k)$ for all $1 \leq k \leq N$ and $[x]_0 = 1$. Thus, the entries of the Vandermonde matrix $\mathbf{X}$ can be expressed in terms of symmetric functions; the lemma below is taken from Oruç and Phillips (2000).

LEMMA 14.1.– *The entries of the Vandermonde matrix* $\mathbf{X} = \mathbf{V}_N(\mathbf{x}) = \{x_j^i\}_{i,j=0}^{N}$ *can be expressed in the form*

$$x_j^i = \sum_{k=0}^{i} h_{i-k}(x_0, x_1, \ldots, x_k)[x_j]_k,\quad i, j = 0, 1, \ldots, N.$$

where $h_{i-k}(x_0, x_1, \ldots, x_k)$ *are the complete symmetric functions and* $[x_j]_k = \prod_{j \neq k}(x_j - x_k)$.

Recalling from the Newton interpolating polynomial $P_N(x)$ for a function $f(x)$ at distinct points $x_0, x_1, \ldots, x_N$ can be expressed in the form (Muhumuza *et al*. 2018a)

$$P_N(x) = f(x_0) + f[x_0, x_1][x]_1 + f[x_0, x_1, x_2][x]_2 + \ldots + f[x_0, x_1, \ldots, x_n][x]_n \tag{14.4}$$

where $[x]_j = (x - x_0)(x - x_1)\cdots(x - x_{j-1})$ and

$$f[x_0, x_1, \ldots, x_j] = \sum_{i=0}^{j} \frac{f(x_i)}{[x_i]_k},\ [x_i]_k = \prod_{k=0, i\neq k}^{j}(x_i - x_k),\ 0 \leq i < j \leq N \tag{14.5}$$

is the jth divided difference of $f(x)$ with respect to the points $x_0, x_1, \ldots, x_N$ and when $f(x)$ is a polynomial of degree at most N, then $P_N(x) = f(x)$. This forms the basis for the **LU** factorization of the Vandermonde matrix and its inverse, as fully discussed in Gautschi (1981) and Gautschi and Inglese (1987). The **LU** can be directly transformed to the **LDU** decomposition, as discussed in Oruç and Phillips (2000) and Oruç and Akmaz (2004):

THEOREM 14.1.– *Given a Vandermonde matrix* $\mathbf{X} \equiv \mathbf{V}_N(\mathbf{x}) = \{x_i^j\}_{i,j=0}^N$ *to be factored into* $\mathbf{X} = \mathbf{LDU}$*, where* $\mathbf{L}$ *is a lower triangular matrix with ones on the major diagonal,* $\mathbf{D}$ *is a diagonal matrix and* $\mathbf{U}$ *is an upper triangular matrix with ones on the major diagonal, the entries of* $\mathbf{L}, \mathbf{D}$ *and* $\mathbf{U}$ *can be expressed by*

$$l_{ij} = \prod_{k=0}^{j-1} \frac{x_i - x_{j-k-1}}{x_j - x_{j-k-1}}, \quad 1 \le j < i \le N, \qquad [14.6]$$

$$d_{ij} = \prod_{k=0}^{j-1} (x_i - x_{j-k-1}), \quad i = j, \qquad [14.7]$$

$$u_{ij} = h_{j-i}(x_1, \ldots, x_i), \quad 1 \le j < i \le N, \qquad [14.8]$$

where the empty product equals one and h_j *are the complete symmetric functions as defined in* [14.3].

The detailed discussion of the statement of proof to some of the results above can be obtained in Oruç and Phillips (2000, 2007) and Oruç and Akmaz (2004).

Based on the above results of **LDU** factorization and applying the general properties of matrices, we directly evaluate the determinant of the Vandermonde matrix **X**, as expressed in the following theorem.

THEOREM 14.2.– *If* $\mathbf{X} = \mathbf{V}_N(\mathbf{x})$ *is a Vandermonde matrix which is diagonalized to* **LDU** *as in Theorem 14.1 above, then the determinant of* **X** *is given by the product of the entries on the major diagonal in* [14.7], *i.e.*

$$\det(\mathbf{X}) = \prod_{j=1}^{N} d_{jj} = \prod_{i=1}^{N} \prod_{k=0}^{j-1} (x_j - x_{j-k-1}) = \prod_{0 \le j < k \le N} (x_j - x_k). \qquad [14.9]$$

PROOF.– If $\mathbf{X} = \mathbf{LDU}$ as in Theorem 14.1, then $\det(\mathbf{X}) = \det(\mathbf{D})$. Since, from [14.7],

$$d_{ij} = \prod_{k=0}^{j-1} (x_i - x_{j-k-1}) \text{ for all } i = j$$

and $\det(\mathbf{D}) = \prod_{j=1}^{N} d_{jj}$, we can write

$$\det(\mathbf{X}) = \det(\mathbf{D}) = \prod_{j=1}^{N} d_{jj} = \prod_{j=1}^{N}\prod_{k=0}^{j-1}(x_j - x_{j-k-1})$$

$$= \prod_{j=1}^{N}\prod_{1\leq i<j}(x_j - x_i) = \prod_{1\leq i<j\leq N}(x_j - x_i).$$

Comparing this to the general expression of the determinant of the Vandermonde matrix, as stated in [14.1], shows that the proof is complete. □

The above matrix decomposition techniques can also be directly applied to the Wishart matrix $\mathbf{W}$, since $\mathbf{W} = \mathbf{X}^\top\mathbf{X}$ (detailed discussion on the same can be found in Yang and Qiao (2003) and Yang (2005, 2007a, b)). In the following sections, we will use this result to show an interesting relation between the extreme points of the Vandermonde determinant and the condition number of the Wishart matrix.

14.4. Matrix norm of the Vandermonde and Wishart matrices

Matrix norms and spectral norms are of great importance in giving bounds for the spectrum of a matrix (von Neumann *et al.* 1963; Gautschi 1990; Muhumuza *et al.* 2019).

The definition of matrix norm $\|\cdot\|$, i.e.

DEFINITION 14.2.– *For a real matrix* $\mathbf{X}$*,*

$$\|\mathbf{X}\|_2 = \sqrt{\rho(\mathbf{X}^\top\mathbf{X})} \leq \sqrt{tr(\mathbf{X}^\top\mathbf{X})} = \|\mathbf{X}\|_F \qquad [14.10]$$

where $\rho(\cdot)$ *denotes the spectral radius,* $tr(\cdot)$ *is the trace,* $\|\cdot\|_2$ *is the natural* L_2*-norm and* $\|\cdot\|_F$ *is the Frobenius norm,*

$$\|\mathbf{X}\|_F = \left(\sum_{i,j} x_{ij}^2\right)^{\frac{1}{2}} \quad \textit{with } \mathbf{X} = [x_{ij}].$$

Using this important definition, we can express the matrix norms of the Wishart matrix. We will also use the following lemma from Muhumuza *et al.* (2018b).

LEMMA 14.2.– *For any symmetric $n \times n$ matrix A with eigenvalues $\{\lambda_i, i = 1, \ldots, n\}$ that are all distinct, and any polynomial P:*

$$\sum_{k=1}^{n} P(\lambda_k) = \mathrm{tr}\left(P(\mathbf{A})\right).$$

PROOF.– By definition, for any eigenvalue λ and eigenvector $\vec{v}$, we must have $\mathbf{A}\vec{v} = \lambda\vec{v}$ and thus

$$P(\mathbf{A})\vec{v} = \left(\sum_{k=0}^{m} c_k \mathbf{A}^k\right)\vec{v} = \sum_{k=0}^{m} c_k (A^k \vec{v}) = \sum_{k=0}^{m} c_k \lambda^k \vec{v}$$

and thus $P(\lambda)$ is an eigenvalue of $P(\mathbf{A})$. For any matrix, $\mathbf{A}$, the sum of eigenvalues is equal to the trace of the matrix

$$\sum_{k=1}^{n} \lambda_k = \mathrm{tr}(\mathbf{A})$$

when multiplicities are taken into account. For the matrices considered in Lemma 14.2, all eigenvalues are distinct. Thus, applying this property to the matrix $P(\mathbf{A})$ gives the desired statement. □

THEOREM 14.3.– *Let $\mathbf{X} = \mathbf{V}_N(\mathbf{x}) = \left\{x_i^j\right\}_{i=0}^{N}$ be a Vandermonde matrix and $\mathbf{W} = (\mathbf{X}^\top \mathbf{X})$ be the Wishart matrix, where $\mathbf{W}$ is diagonalizable. Then, the matrix norm of $\mathbf{W}$ can be expressed as*

$$\|\mathbf{W}\|_F^2 = \sum_{j=1}^{N} \eta_j^4 = \mathrm{tr}(\mathbf{W}^2) \qquad [14.11]$$

and

$$\|\mathbf{W}^{-1}\|_F^2 = \det(\mathbf{X})^{-4} \cdot \sum_{j=1}^{N} \left(\prod_{\substack{k=1 \\ i \neq k}}^{N} \eta_k^4 \right) \qquad [14.12]$$

where $\|\cdot\|_F$ is the Frobenius norm and η_j's are the singular values $\mathbf{X}$.

PROOF.– Let $\mathbf{X} \in \mathcal{M}_N$. Then, by singular value decomposition (SVD), i.e. by definition, the matrix $\mathbf{X}$ can be expressed as

$$\mathbf{X} = \mathbf{UDV} \qquad [14.13]$$

where $\mathbf{U} = [u_1, \ldots, u_N]$ is an $N \times N$ orthogonal matrix and $\mathbf{V} = [v_1, \ldots, v_N]$ is an $N \times N$ orthogonal matrix, $\mathbf{D} = \text{diag}(\eta_1, \ldots, \eta_N)$ is an $N \times N$ diagonal matrix and $\mathbf{U}^\top\mathbf{U} = \mathbf{V}^\top\mathbf{V} = \mathbf{I}$. It follows that

$$\mathbf{X}^\top\mathbf{X} = \mathbf{V}^\top\mathbf{D}\mathbf{U}^\top\mathbf{U}\mathbf{D}\mathbf{V} = \mathbf{V}^\top\mathbf{D}^2\mathbf{V},\ \mathbf{U}^\top\mathbf{U} = \mathbf{I}.$$

Using the definition of the matrix norm of $\mathbf{X}$, we can then write

$$\|\mathbf{X}\|_F^2 = \sum_{i=1}^{N}\sum_{j=1}^{N} x_{ij}^2 = \sum_{i=1}^{N}\sum_{j=1}^{N} \eta_j^2 v_{ji}^2. \qquad [14.14]$$

Since $\mathbf{V}$ is an orthogonal matrix, each column must have unit length, in other words, $\sum_{j=1}^{N} v_{ji}^2 = 1$. Thus, we can express the norm as follows:

$$\|\mathbf{X}\|_F^2 = \sum_{i=1}^{N}\sum_{j=1}^{N} x_{ij}^2 = \sum_{i=1}^{N} \eta_j^2 \sum_{j=1}^{N} v_{ji}^2 = \sum_{i=1}^{N} \eta_j^2.$$

Since $\mathbf{W} = \mathbf{X}^\top\mathbf{X}$ and applying SVD [14.13], following the steps above and the fact that $\|\mathbf{W}\|_F = \|\mathbf{X}^\top\mathbf{X}\|_F \leq \|\mathbf{X}^\top\|_F\|\mathbf{X}\|_F$, we have

$$\|\mathbf{W}\|_F^2 = \|\left(\mathbf{V}\mathbf{D}^4\mathbf{V}^T\right)\|_F^2 = \sum_{j=1}^{N} \eta_j^4 \sum_{j=1}^{N} |v_{ji}|^2 = \sum_{j=1}^{N} \eta_j^4. \qquad [14.15]$$

Here we note that $\mathbf{W}$ is a symmetric matrix so by Lemma 14.2

$$\text{tr}(\mathbf{W}^2) = \sum_{j=1}^{N} \lambda_j^2$$

where λ_i are the eigenvalues of $\mathbf{W}$ and by the SVD $\mathbf{W} = \mathbf{V}^\top\mathbf{D}^2\mathbf{V}$, we can conclude that $\lambda_i = \eta_i^2$, $\sum_{j=1}^{N} \eta_j^4 = \text{tr}(\mathbf{W}^2)$ and thus

$$\|\mathbf{W}\|_F = \sum_{j=1}^{N} \eta_j^4 = \text{tr}(\mathbf{W}^2).$$

Next, we will prove the expression for the norm of $\mathbf{W}^{-1}$. Since $\mathbf{U}$ is a unitary matrix

$$\|\mathbf{W}^{-1}\|_F = \|\mathbf{U}\mathbf{D}^{-2}\mathbf{U}^\top\|_F = \|\mathbf{U}\|_F\|\mathbf{D}^{-2}\|_F\|\mathbf{U}^\top\|_F = \|\mathbf{D}^{-2}\|_F = \text{tr}(\mathbf{D}^{-4}).$$

Since $\mathbf{D}$ is a diagonal matrix, it can be rewritten as $\mathbf{D} = \text{diag}(\eta_1, \eta_2, \eta_3, \cdots, \eta_N)$. It follows that

$$\mathbf{D}^{-1} = \begin{bmatrix} \frac{1}{\eta_1} & 0 & 0 & \cdots & 0 \\ 0 & \frac{1}{\eta_2} & 0 & \cdots & 0 \\ 0 & 0 & \frac{1}{\eta_3} & \cdots & 0 \\ \vdots & \vdots & \vdots & \ddots & \vdots \\ 0 & 0 & 0 & \cdots & \frac{1}{\eta_N} \end{bmatrix} = \frac{1}{\prod_{i=1}^{N} \eta_i} \begin{bmatrix} \prod_{i\neq 1, i=1}^{N} \eta_i & 0 & \cdots & 0 \\ 0 & \prod_{i\neq 2, i=1}^{N} \eta_i & \cdots & 0 \\ \vdots & \vdots & \ddots & \vdots \\ 0 & 0 & \cdots & \prod_{i\neq N, i=1}^{N} \eta_i \end{bmatrix}.$$

Thus, $\text{tr}(\mathbf{D}^{-4}) = \sum_{j=1}^{N} \tilde{\eta}_j^4$, where $\tilde{\eta}_j = \prod_{\substack{k=1 \\ i\neq k}}^{N} \eta_k \Big/ \prod_{i=1}^{N} \eta_i$. Since $\mathbf{D}$ is a diagonal matrix, then $\det(\mathbf{D}) = \prod_{j=1} \eta_j$ and since

$$\det(\mathbf{X}) = \det(\mathbf{U}^\top \mathbf{D} \mathbf{V}) = \det(\mathbf{U}^\top) \det(\mathbf{D}) \det(\mathbf{V}) = \det(\mathbf{D})$$

then $\tilde{\eta}_j = \prod_{\substack{k=1 \\ i\neq k}}^{N} \eta_k \Big/ \prod_{i=1}^{N} \eta_i = \prod_{\substack{k=1 \\ i\neq k}}^{N} \eta_k \Big/ \det(\mathbf{X})$. Combining this with the expression for $\text{tr}(\mathbf{D}^{-4})$ above, we get the desired expression for the norm of the inverse. □

The concept of the matrix norm is closely related to the condition number that we directly apply to the case of the Vandermonde matrix and the Wishart matrix using extreme points, as discussed in the next section.

14.5. Condition number of the Vandermonde and Wishart matrices

The concept of the condition number for the Vandermonde matrix has previously been investigated in connection with the stability of polynomial interpolation and least squares as discussed in Cheney (1966), Beckermann and Labahn (2000), Dunkl and Xu (2001), Beckermann *et al.* (2007), Kaltofen *et al.* (2012) and von Neumann *et al.* (1963), Smale (1985), Demmel (1988, 1997), Dubiner (1991), Brutman (1997) and Calvi and Levenberg (2008), (Bos *et al.* 2010a, 2010b).

Our aim in this section is to illustrate the significance of the extreme points of the Vandermonde matrix in order to optimize its condition number as a measure of sensitivity and stability of the given system. Condition numbers can explain the best possible accuracy of the solution of say, a linear system, i.e. $\mathbf{Ax} = \mathbf{b}$ in the presence of approximations made by the computation (Bos *et al.* 2010a, 2010b). The condition number can also bound the rate of convergence of iterative methods, measure distance of an instance to singularity and/or shed light on preconditioning (Hoel 1958; Golub and Van Loan 1996; Higham 2002).

DEFINITION 14.3.– *The conditional number of a matrix say,* $\mathbf{X}$ *denoted by cond*$(\mathbf{X})$ *or* $\kappa(\mathbf{X})$ *is defined as*

$$\kappa(\mathbf{X}) = \|\mathbf{X}\|\|\mathbf{X}^{-1}\| \qquad [14.16]$$

Since the condition number of a matrix is expressed in terms of the norm of the matrix and the norm of its inverse, then we can express the condition number in terms of the matrix trace and determinant. We demonstrate this for the case of the Vandermonde matrix and the Wishart.

THEOREM 14.4.– *Let* $\mathbf{X} = \mathbf{V}_N(\mathbf{x}) = \left\{x_i^j\right\}_{i=0}^{N}$ *be a Vandermonde matrix, and* $\mathbf{W} = (\mathbf{X}^\top\mathbf{X})$ *be the Wishart matrix. Then, the conditional number of* $\mathbf{W}$, $\kappa(\mathbf{W})$ *can be minimized by maximizing the determinant of* $\mathbf{X}$.

PROOF.– Applying Theorem 14.3 in the previous section, and also the ideas of the Vandermonde determinant, then we can express the matrix norm in [14.11] and [14.12] and since $\mathbf{X}$ is the Vandermonde matrix, then it follows that

$$\|\mathbf{X}\|_F^2 = \mathrm{tr}(\mathbf{X}^\top\mathbf{X}) = \sum_{j=1}^{N} \eta_j^2, \text{ and } \|\mathbf{X}^{-1}\|_F^2 = \det(\mathbf{X})^{-2} \cdot \sum_{j=1}^{N} \tilde{\eta}_j^2$$

where $\tilde{\eta}_j = \prod_{\substack{k=1 \\ i \neq k}}^{N} \eta_k \Big/ \prod_{i=1}^{N} \eta_i = \prod_{\substack{k=1 \\ i \neq k}}^{N} \eta_k \Big/ \det(\mathbf{X})$.

Thus, the condition number of $\mathbf{X}$, $\kappa(\mathbf{X})$ will be given by

$$\kappa(\mathbf{X}) = \|\mathbf{X}\|_F\|\mathbf{X}^{-1}\|_F = \det(\mathbf{X})^{-1} \cdot \left(\sum_{j=1}^{N} \eta_j^2\right)^{\frac{1}{2}} \cdot \left(\sum_{j=1}^{N} \tilde{\eta}_j^2\right)^{\frac{1}{2}} \qquad [14.17]$$

Since the product term in the denominator, which happens to be the determinant of the Vandermonde matrix, is a dominant term compared to the partial sums in the numerator, then the value of the condition number, $\kappa(\mathbf{X})$, can be minimized by maximizing the Vandermonde determinant.

Similarly, since the Wishart matrix $\mathbf{W} = \mathbf{X}^\top \mathbf{X}$, then its condition number $\kappa(\mathbf{W})$ follows immediately from the fact that

$$\|\mathbf{W}\|_F^2 = \sum_{j=1}^{N} \eta_j^4,$$

and

$$\|\mathbf{W}^{-1}\|_F^2 = \det(\mathbf{X})^{-4} \cdot \sum_{j=1}^{N} \tilde{\eta}_j^4.$$

Therefore, we obtain the condition number $\kappa(\mathbf{W})$ of the Wishart matrix as

$$\kappa(\mathbf{W}) = \|\mathbf{W}\|_F \|\mathbf{W}^{-1}\|_F = \det(\mathbf{X})^{-2} \cdot \mathrm{tr}(\mathbf{W}^2)^{\frac{1}{2}} \cdot \left(\sum_{j=1}^{N} \tilde{\eta}_j^4\right)^{\frac{1}{2}} \quad [14.18]$$

The term $\det(\mathbf{X})$ in the denominator is the Vandermonde determinant, and increasing its value will often make the condition numbers of $\mathbf{X}$ and $\mathbf{W}$ smaller. □

	x_1	x_2	x_3	λ_1	λ_2	λ_3
$\mathbf{p}_{\max}$	-0.7071	0.0000	0.7071	3.351	0.1492	1.0000
$\mathbf{p}_1$	0.9212	-0.3859	0.0489	3.631	1.0468	0.06467
$\mathbf{p}_2$	0.6932	-0.6080	0.3877	3.266	1.5824	0.02217
$\mathbf{p}_3$	0.7871	-0.2923	.5432	3.845	0.6127	0.02054
$\mathbf{p}_4$	0.9873	0.1561	-8.2000×10^{-3}	3.976	0.4866	0.06057
$\mathbf{p}_5$	0.9428	0.3155	0.1068	4.270	0.3119	0.01786
$\mathbf{p}_{\min}$	0.9990	0.0445	-0.9500×10^{-3}	4.021	0.9741	6.89×10^{-4}

Table 14.1. *For different points on a three-dimensional sphere and the square of the eigenvalues of the corresponding Vandermonde matrix. Here, $\mathbf{p}_{\max}$ is a point that maximizes the Vandermonde determinant and $\lambda_i = \eta_i^2$, where η_i is the ith eigenvalue of the Vandermonde matrix*

To illustrate the above results, we chose a few points on the sphere and computed the condition number using [14.17] and [14.18], as well as the definition. The results are shown in Tables 14.1 and 14.2. From these tables, we note that the condition number of both the Vandermonde matrix $\kappa(\mathbf{X})$ and the Wishart matrix $\kappa(\mathbf{W})$, with respect to Frobenius norm, is highly dependent and sensitive to the Vandermonde determinant $|\mathbf{X}|$. That is, $\kappa(\mathbf{X}) \propto |\mathbf{X}|^{-1}$ and $\kappa(\mathbf{W}) \propto |\mathbf{X}|^{-2}$ or simply $\kappa(\mathbf{W}) \propto \kappa(\mathbf{X})^2$. Therefore, the extreme points that maximize the Vandermonde determinant can minimize the condition number. These points are often referred to as Fekete points, as discussed in Muhumuza *et al*. (2018a). In the next section, we demonstrate that these extreme points are actually the eigenvalues of the Wishart matrix whose joint eigenvalue probability density function is a Gaussian ensemble, as discussed in Muhumuza *et al*. (2018b).

		$\kappa(\mathbf{X})_F$		$\kappa(\mathbf{W})_F$	
	$\lvert\mathbf{X}\rvert$	Using (14.17)	Using definition	Using (14.18)	Using definition
$\mathbf{p}_{max}$	0.7071	6.000	6.000	23.74	23.74
$\mathbf{p}_1$	0.4958	8.898	8.898	58.56	58.56
$\mathbf{p}_2$	0.3958	12.81	14.97	119.7	163.7
$\mathbf{p}_3$	0.1360	23.23	9.226	422.7	66.66
$\mathbf{p}_4$	0.1094	23.30	16.53	476.8	240.0
$\mathbf{p}_5$	0.0739	37.24	37.24	1257	1257
$\mathbf{p}_{min}$	0.05198	85.14	85.14	5999	5999

Table 14.2. *Comparison of the value of the Vandermonde determinant ($\lvert\mathbf{X}\rvert$) and the condition number of the corresponding* $\mathbf{X}$ *and* $\mathbf{W}$ *for the points given in Table 14.1*

14.6. Conclusion

In this study, we establish the usefulness of the polynomial decomposition of the properties of the extreme points of the joint eigenvalue probability density function, which are helpful in the optimization of the condition number of the Vandermonde and Wishart matrices by maximizing the Vandermonde determinant. This is justified by the fact that the condition number of the Vandermonde matrix is inversely proportional to the absolute value of its determinant, while the condition of the Wishart matrix is inversely proportional to the square of the Vandermonde determinant. Therefore, the extreme points of the joint probability density function of the Wishart matrix that maximize the Vandermonde determinant can be used to minimize the condition number of both the Vandermonde determinant and the Wishart matrix. The points that maximize the Vandermonde determinant are often referred to as Fekete points.

We have also been able to illustrate that the extreme points of the Vandermonde determinant are indeed related to the eigenvalues of the Wishart and these extreme points have a joint eigenvalue density function that is a Gaussian ensemble. These points, which are also zero of classical orthogonal polynomials, provide the most stable and economical interpolating points.

Our future plan is to apply these results to the optimal control theory, especially in finance and high dimensional data analysis.

14.7. Acknowledgments

We acknowledge the financial support for this research by the Swedish International Development Agency (Sida), Grant No. 316, International Science Program (ISP) in Mathematical Sciences (IPMS). We are also grateful to the Division of Applied Mathematics, Mälardalen University, for providing an excellent and inspiring environment for education and research.

14.8. References

Abramowitz, M. and Stegun, I.A. (1965). *Handbook of Mathematical Functions: with Formulas, Graphs, and Mathematical Tables 55*, Courier Corporation.

Anderson, T.W. (2003). *An Introduction to Multivariate Statistical Analysis*, 3rd edition. John Wiley & Sons, Inc., Hoboken, New Jersey.

Beckermann, B. and Labahn, G. (2000). Effective computation of rational approximants and interpolants. *Reliable Computing*, 6(4), 365–390.

Beckermann, B., Golub, G.H., Labahn, G. (2007). On the numerical condition of a generalized Hankel eigenvalue problem. *Numerische Mathematik*, 106(1), 41–68.

Bjorck, A. and Elfving, T. (1973). Algorithms for confluent Vandermonde systems. *Numer. Math.*, 21, 130–137.

Bjorck, A. and Pereyra, V. (1970). Solution of Vandermonde systems of equations. *Math. Comp.*, 24, 893–903.

Bos, L., Sommariva, A., Vianello, M. (2010a). Least-squares polynomial approximation on weakly admissible meshes: Disk and triangle. *Journal of Computational and Applied Mathematics*, 235(3), 660–668.

Bos, L., Calvi, J.-P., Levenberg, N., Sommariva, A., Vianello, M. (2010b). Geometric weakly admissible meshes, discrete least squares approximation and approximate Fekete points. *Math. Comp.*

Brutman, L. (1997). Lebesgue functions for polynomial interpolation – A survey. *Ann. Numer. Math.*, 4, 111–127.

Calvi, J.P. and Levenberg, N. (2008). Uniform approximation by discrete least squares polynomials. *J. Approx. Theory*, 152, 82–100.

Cheney, E.W. (1966). *Introduction to Approximation Theory*. McGraw-Hill, New York.

Demmel, J.W. (1988). The probability that a numerical analysis problem is difficult. *Mathematics of Computation*, 50(182), 449–480.

Demmel, J.W. (1997). *Applied Numerical Linear Algebra 56*. SIAM, Philadelphia.

Dubiner, M. (1991). Spectral methods on triangles and other domains. *J. Sci. Comput.*, 6, 345–390.

Dunkl, C.F. and Xu, Y. (2001). *Orthogonal Polynomials of Several Variables*. Cambridge University Press, Cambridge.

Dyson, F.J. (1962). Statistical theory of the energy levels of complex systems. *Journal of Mathematical Physics*, 3(1), 140–156.

Edelman, A. (1988). Eigenvalues and condition numbers of random matrices. *SIAM Journal on Matrix Analysis and Applications*, 9(4), 543–560.

El-Mikkawy, M. (2003). Explicit inverse of a generalized Vandermonde matrix. *Appl. Math. Comput.*, 146, 643–652.

El-Mikkawy, M. and El-Desouky, B. (2003). On a connection between symmetric polynomials, generalized Stirling numbers and the Newton general divided difference interpolation polynomial. *Appl. Math. Comput.*, 138, 375–385.

Forester, P.J. (2010). *Log–Gases and Random Matrices*. London Mathematical Society Monographs, Princeton University Press, London.

Gautschi, W. (1981). A survey of Gauss–Christoffel quadrature formulae. *EB Christoffel*, Birkhäuser, Basel.

Gautschi, W. (1990). How (stable) are Vandermonde systems. *Asymptotic and Computational Analysis*, 124, 193–210.

Gautschi, W. and Inglese, G. (1987). Lower bounds for the condition number of Vandermonde matrices. *Numerische Mathematik*, 52(3), 241–250.

Golub, G.H. and Van Loan, C.F. (1996). *Matrix Computations*, 3rd edition. Johns Hopkins University Press, Baltimore.

Higham, N.J. (2002). *Accuracy and Stability of Numerical Algorithms*, 2nd edition. SIAM, Philadelphia.

Hoel, P.G. (1958). Efficiency problems in polynomial estimation. *The Annals of Mathematical Statistics*, 29(4), 1134–1145.

Kaltofen, E.L., Lee, W.S., Yang, Z. (2012). Fast estimates of Hankel matrix condition numbers and numeric sparse interpolation. *Proceedings of the 2011 International Workshop on Symbolic-Numeric Computation*, 130–136.

Karlin, S., and Studden, W.J. (1966). *Tchebycheff System: With Applications in Analysis and Statistics*. Interscience Publishers, John Wiley & Sons, New York.

König, W. (2005). Orthogonal polynomial ensembles in probability theory. *Probability Surveys*, 2, 385–447.

Ljung, L., Pflug, G., Harro, W. (2012). *Applied Stochastic Approximation and Optimization of Random Systems 17*. Birkhäuser, Basel.

Lundengård, K., Österberg, J., Silvestrov, S. (2013). Extreme points of the Vandermonde determinant on the sphere and some limits involving the generalized Vandermonde determinant. arXiv, eprint arXiv:1312.6193.

Macdonald, I.G. (1979). *Symmetric Functions and Hall Polynomials*. Oxford University Press, Oxford.

Martinez, J.J. and Pena, J.M. (1998). Factorization of Cauchy–Vandermonde matrices. *Linear Algebra Appl.*, 284, 229–237.

Mehta, M.L. (1967). *Random Matrices and the Statistical Theory of Energy Levels*. Academic Press, New York, London.

Muhumuza, A.K., Lundengård, K., Österberg, J., Silvestrov, S., Mango, J.M., Kakuba, G. (2018a). The generalized Vandermonde interpolation polynomial based on divided differences. In *Proceedings of the SMTDA2018 Conference*, Skiadas, C.H. (ed.). Crete.

Muhumuza, A.K., Lundengård, K., Österberg, J., Silvestrov, S., Mango, J.M., Kakuba, G. (2018b). The multivariate Wishart distribution based on generalized Vandermonde determinant. Submission, Methodology and Computing in Applied Probability, IWAP2018 Conference.

Muhumuza, A.K., Lundengård, K., Österberg, J., Silvestrov, S., Mango, J.M., Kakuba, G. (2019). Notes on the extreme points of the Vandermonde determinant on surfaces implicitly determined by a univariate polynomial. In *Algebraic Structures and Applications. SPAS2017, Västerås and Stockholm, Sweden, October 4 – 6, 2017*, Silvestrov, S., Malyarenko, A., Rančić, M. (eds). Springer, Cham.

von Neumann, J., Taub, A.W., Taub, A.H. (1963). *The Collected Works of John von Neumann: 6-Volume Set*. Reader's Digest Young Families, Pleasantville, New York.

Oruç, H. and Akmaz, H.K. (2004). Symmetric functions and the Vandermonde matrix. *J. Comput. Appl. Math.*, 172, 49–64.

Oruç, H. and Phillips, G.M. (2000). Explicit factorization of the Vandermonde matrix. *Linear Algebra Appl.*, 315, 113–123.

Oruç, H. and Phillips, G.M. (2007). LU factorization of the Vandermonde matrix and its applications. *Appl. Math. Lett.*, 20, 892–897.

Phillips, G.M. (2003). *Interpolation and Approximation by Polynomials*. Springer-Verlag, New York.

Smale, S. (1985). On the efficiency of algorithms of analysis. *Bull. New. Ser. Am. Math. Soc.*, 13(2), 87–121.

Spivey, M.Z. and Zimmer, A.M. (2008). Symmetric polynomials, Pascal matrices, and Stirling matrices. *Linear Algebra Appl.*, 428(4), 1127–1134.

Szegő, G. (1939). *Orthogonal Polynomials*. American Mathematics Society, Rhode Island.

Tang, W.P. and Golub, G.H. (1981). The block decomposition of a Vandermonde matrix and its applications. *BIT*, 21, 505–517.

Wigner, E.P. (1951). On the statistical distribution of the widths and spacings of nuclear resonance levels. *Mathematical Proceedings of the Cambridge Philosophical Society*, 7(4).

Yang, S.-L. (2005). On the LU factorization of the Vandermonde matrix. *Discrete Appl. Math.*, 146(2), 102–105.

Yang, S.-L. (2007a). On a connection between the Pascal, Stirling and Vandermonde matrices. *Discrete Appl. Math.*, 155(2), 2025–2030.

Yang, S.-L. (2007b). Generalized Leibniz functional matrices and factorization of some well-known matrices. *Linear Algebra Appl.*, 430(1), 511–531.

Yang, S.-L. and Qiao, Z.-K. (2003). Stirling matrix and its property. *Int. J. Appl. Math.*, 14(2), 145–157.

15

Forecast Uncertainty of the Weighted TAR Predictor

In this chapter, we investigate the forecast uncertainty of a new predictor proposed for the Self-Exciting Threshold AutoRegressive (SETAR) model. We consider a weighted mean predictor whose weights are obtained from the minimization of the Mean Square Forecast Errors (MSFE). Even though the "point accuracy" of this predictor has been investigated, the study of its distribution and, in particular, the construction of the prediction intervals have not been studied. Starting from the evaluation that the predictor follows a nonstandard distribution, in this chapter, we focus the attention on the generation of prediction intervals for the weighted SETAR predictor, using different bootstrap methods for dependent data. The coverage and the length of the prediction intervals are evaluated and compared through a Monte Carlo study.

15.1. Introduction

In time series analysis, the forecast generation is often confined to the point forecasts that undoubtedly have high relevance from the empirical point of view, but do not give any information on the uncertainty of the predictor. In fact, the point forecasts are usually evaluated considering indices that give evidence of how the predicted value is "far" from the observed data or, in other cases, how the predictor makes it possible to obtain forecasts that are more (or less) accurate than those obtained from other predictors, but no information is given on their likely accuracy[1].

Chapter written by Francesco GIORDANO and Marcella NIGLIO.

1 In this chapter, by *forecast* (or *prediction*) we mean the value obtained from a given predictor; whereas the *predictor* is the function (obtained from the "optimization" of a well-defined loss function) that, given the estimated parameters and the past values, makes it possible to generate forecasts.

In this chapter, we focus the attention on the forecast uncertainty of a point predictor which is assessed defining an interval in which the prediction belongs with a given probability. In other words, our aim is to define a prediction interval (called a PI in the following).

In more detail, we consider the PI of a new predictor proposed for the widely known nonlinear time series model: the Threshold AutoRegressive model (see Tong 1978, 1983, 1990) whose structure is shortly presented here.

Let X_t, with $t \in T$, be a nonlinear Threshold AutoRegressive (TAR) model; its stochastic structure is given by

$$X_t = \sum_{j=1}^{k} \left(\phi_0^{(j)} + \phi_1^{(j)} X_{t-1} + \ldots + \phi_{p_j}^{(j)} X_{t-p_j} \right) \mathbb{I}_{\{Y_{t-d} \in \mathbb{R}_j\}} + \epsilon_t, \qquad [15.1]$$

where k is the number of regimes, p_j is the autoregressive order of regime j, $\mathbb{I}_{\{\cdot\}}$ is the indicator function, Y_{t-d} is the threshold variable, d is the threshold delay, $\mathbb{R}_j$ is the subset of the real line such that $\mathbb{R}_j \cap \mathbb{R}_{j'} = \emptyset$, for $j \neq j'$ and $\bigcup_{j=1}^{k} \mathbb{R}_j = \mathbb{R}$, with $\mathbb{R}_j = (r_{j-1}, r_j]$ and $-\infty = r_0 < r_1 < \ldots < r_{k-1} < r_k = \infty$, with r_j being the so-called threshold value, for $j = 1, \ldots, k$.

In this large class of nonlinear models, we consider the so-called Self-Exciting TAR (SETAR) model characterized by an endogenous threshold variable, $Y_{t-d} = X_{t-d}$, given by the delayed value of X_t itself.

The generation of point forecasts from this class of models has been largely investigated and different predictors have been proposed in the literature (see, for example, Clements and Smith (1997), Clements *et al.* (2003) and Boero and Marrocu (2004)).

In this context, a new proposal has been given in (Niglio 2019) that introduces a predictor based on the weighted average of past observations and whose weights are obtained from the minimization of the Mean Square Forecast Errors (MSFE).

In more detail, starting from model [15.1], in the following we consider the SETAR parametrization with two regimes ($k = 2$) that we shortly denote as SETAR(2; p, p)

$$\begin{aligned} X_t = &\left(\phi_0^{(1)} + \phi_1^{(1)} X_{t-1} + \ldots + \phi_p^{(1)} X_{t-p} \right) \mathbb{I}_{\{X_{t-d} \in \mathbb{R}_1\}} \\ &+ \left(\phi_0^{(2)} + \phi_1^{(2)} X_{t-1} + \ldots + \phi_p^{(2)} X_{t-p} \right) (1 - \mathbb{I}_{\{X_{t-d} \in \mathbb{R}_1\}}) + \epsilon_t \qquad [15.2] \end{aligned}$$

that, to simplify the presentation of the predictor, is differently represented, in Niglio (2019), with

$$X_t = \mathbb{I}_{t-d}^{\top} \mathbf{\Phi} \mathbf{X}_{t-1} + \epsilon_t, \qquad [15.3]$$

where

$$\mathbb{I}_{t-d} = \begin{pmatrix} \mathbb{I}_{\{X_{t-d}\in\mathbb{R}_1\}} \\ 1-\mathbb{I}_{\{X_{t-d}\in\mathbb{R}_1\}} \end{pmatrix}, \quad \mathbf{\Phi} = \begin{pmatrix} \phi_0^{(1)} & \phi_1^{(1)} & \dots & \phi_p^{(1)} \\ \phi_0^{(2)} & \phi_1^{(2)} & \dots & \phi_p^{(2)} \end{pmatrix}, \quad \mathbf{X}_{t-1} = \begin{pmatrix} 1 \\ X_{t-1} \\ X_{t-2} \\ \dots \\ X_{t-p} \end{pmatrix}$$

and $\mathbb{I}_{t-d}^{\top}$ is the transpose of the vector $\mathbb{I}_{t-d}$.

Given the parametrization [15.3], the weighted one-step-ahead predictor of X_{T+1} is given by

$$\hat{X}_{T+1}^{w} = \sum_{t=\max\{p,d\}+1}^{T} w_t X_t = \sum_{t=\max\{p,d\}+1}^{T} w_t(\mathbb{I}_{t-d}^{\top}\mathbf{\Phi}\mathbf{X}_{t-1} + \epsilon_t), \qquad [15.4]$$

where the weights w_t, $t = \max\{p, d\} + 1, \dots, T$, are obtained from the minimization of the Mean Square Forecast Error

$$\min_{\boldsymbol{w}\in\boldsymbol{W}} E[(X_{T+1} - \hat{X}_{T+1}^{w})^2],$$

with $\mathbf{W}$ being a compact set for $\mathbf{w}$.

Niglio (2019) shows that the main advantage in the predictor [15.4] is that it takes into account all the observed data (and then the whole data generating process) differently from other predictors, such as the conditional expectations, which only involve the last observed values. From the computational point of view, the minimization and, therefore, the derivation of the weights is not an easy task, but when the forecasting performance of [15.4] is compared to the SETAR predictor obtained from the conditional expectation of X_{T+1} or to other linear predictors (such as the AR(p) or the Random Walk predictors), it often outperforms its competitors. In particular, it is especially evident when the nonlinearity and the persistence of the generating process grow and when the difference of the parameters between the two regimes increases. On the other hand, when the number of observations that belong to a single regime is limited and it is difficult to discriminate between a linear AR and a nonlinear SETAR model, the weighted predictor does not outperform the linear competitor.

The weighted predictor [15.4] has been discussed and evaluated in Niglio (2019); however, its distribution and, in particular, the assessment of its uncertainty have not been faced.

Starting from this point, in this chapter, we evaluate the uncertainty of the predictor [15.4] through the generation of PI's. In section 15.2, in more detail, we introduce the main results given in the literature on the generation of bootstrap PI's in the SETAR domain and we adapt those approaches to the weighted predictor constructed for X_{T+1}. The theoretical details given in section 15.2 are then evaluated in section 15.3 where the coverage and the length of the PI are investigated through a Monte Carlo study.

15.2. SETAR predictors and bootstrap prediction intervals

The generation of the PI's for SETAR models has often raised the interest of the literature (for recent contributions, see Li (2011) and Staszewska-Bystrova and Winker (2016)).

The nonlinear dynamic of the SETAR model, the variability caused by the parameter estimation and even their bias in the presence of small samples make the distribution of the predictor "not standard". It has led to the use of computational intensive techniques that make it possible to construct PI's even in the presence of complex nonlinear structures.

Starting from this evaluation, Li (2011) and Staszewska-Bystrova and Winker (2016) have used bootstrap approaches to generate PI's whose empirical coverage has been investigated through Monte Carlo studies.

In both contributions, the proposed PI's are related to the SETAR predictor, obtained as the conditional expected value of X_{T+1}, $X^{ce}_{T+1} = E[X_{T+1}|\mathcal{F}_t]$, with $\mathcal{F}_t$ being the set of information available until time T, that, in the presence of model [15.2], with $d = 1$, becomes:

$$X^{ce}_{T+1} = \left(\phi_0^{(1)} + \phi_1^{(1)} X_T + \ldots + \phi_p^{(1)} X_{T-p+1}\right) \mathbb{I}_{\{X_T \in \mathbb{R}_1\}} + \left(\phi_0^{(2)} + \phi_1^{(2)} X_T + \ldots + \phi_p^{(2)} X_{T-p+1}\right) (1 - \mathbb{I}_{\{X_T \in \mathbb{R}_1\}}), \qquad [15.5]$$

where "*ce*" stands for the *c*onditional *e*xpectation.

Among the bootstrap approaches developed in the time series context (for two interesting reviews, see Bühlmann (2002) and Kreiss and Paparoditis (2011)), Li (2011) and Staszewska-Bystrova and Winker (2016) make use of the *residual bootstrap* and define proper procedures that are detailed here and then modified to take into account the predictor [15.4].

The steps of method 3 proposed in Li (2011) are the following:

L1. estimates the parameters of the SETAR model [15.2], following the three steps procedure described in Tong (1983);

L2. computes the fitted values $\hat{\epsilon}_t = X_t - \hat{X}_t$, for $t = p+1, \ldots, T$ and generates the bootstrap replicate X_t^*

$$X_t^* = \left(\hat{\phi}_0^{(1)} + \hat{\phi}_1^{(1)} X_{t-1}^* + \ldots + \hat{\phi}_p^{(1)} X_{t-p}^*\right) \mathbb{I}_{\{X_{t-1}^* \in \hat{\mathbb{R}}_1\}} + \left(\hat{\phi}_0^{(2)} + \hat{\phi}_1^{(2)} X_{t-1}^* + \ldots + \hat{\phi}_p^{(2)} X_{t-p}^*\right) (1 - \mathbb{I}_{\{X_{t-1}^* \in \hat{\mathbb{R}}_1\}}) + \epsilon_t^* \quad [15.6]$$

where $X_t^* = X_t$, for $t \leq p$, ϵ_t^* is drawn with replacement from the $\hat{\epsilon}_t$'s and $\hat{\mathbb{R}}_1$ is the real subset $(-\infty, \hat{r}_1]$.

L3. estimates the model parameters using the bootstrap series X_t^* and so computes the one-step-ahead forecast $\hat{X}_{T+1}^{ce*}$ fixing $X_t^* = X_t$, for $t = T-p+1, \ldots, T$

$$\hat{X}_{T+1}^{ce*} = \left(\hat{\phi}_0^{(1)*} + \hat{\phi}_1^{(1)*} X_T^* + \ldots + \hat{\phi}_p^{(1)*} X_{T-p+1}^*\right) \mathbb{I}_{\{X_T^* \in \hat{\mathbb{R}}_1^*\}} + \left(\hat{\phi}_0^{(2)*} + \hat{\phi}_1^{(2)*} X_T^* + \ldots + \hat{\phi}_p^{(2)*} X_{T-p+1}^*\right) (1 - \mathbb{I}_{\{X_T^* \in \hat{\mathbb{R}}_1^*\}}) + \epsilon_{T+1}^* \quad [15.7]$$

where, even in this case, ϵ_{T+1}^* is randomly drawn from the $\hat{\epsilon}_t$'s and $\hat{\mathbb{R}}_1^*$ is the subset $(-\infty, \hat{r}_1^*]$;

L4. repeats B times the steps L2 and L3, so obtaining the set of bootstrap forecasts $\{\hat{X}_{T+1,1}^{ce*}, \hat{X}_{T+1,2}^{ce*}, \ldots, \hat{X}_{T+1,B}^{ce*}\}$; the $(1-\alpha)\%$ PI is given by $[\hat{X}_{T+1}^{ce*}(\alpha/2), \hat{X}_{T+1}^{ce*}(1-\alpha/2)]$, with $\hat{X}_{T+1}^{ce*}(\alpha/2)$ and $\hat{X}_{T+1}^{ce*}(1-\alpha/2)$ being the $\alpha/2$ and $1-\alpha/2$ quantiles, respectively, of the empirical distribution of the B bootstrap forecasts.

Staszewska-Bystrova and Winker (2016) propose a modification with respect to what is proposed in Li (2011) in step L3. In order to include the assumption given on the innovation term of the SETAR model in the bootstrap procedure, Staszewska-Bystrova and Winker (2016) suggest drawing ϵ_{T+1}^* from the assumed distribution of ϵ_t. They clarify that it makes it possible to consider an ϵ_{T+1}^* coherent with the DGP and they show, through a Monte Carlo simulation, that it guarantees a better coverage of the PI if compared to the approach of Li (2011).

Unfortunately, the PI's obtained from the residual bootstrap approach of Li (2011) and Staszewska-Bystrova and Winker (2016) suffer clear under-coverage in the presence of one-step-ahead predictors.

In this context, another recent contribution is given by Pan and Politis (2016) that faces the generation of PI's for autoregressive (linear, nonlinear and nonparametric) process, considering a six-step procedure that they detail, in the parametric case, for the linear autoregressive models. In the following, Pan and Politis' bootstrap procedure (2016) is detailed for SETAR models and, to give more emphasis on the content of each step, the procedure is expanded to 10 steps.

Note that Pan and Politis (2016) consider, for linear autoregressive models, forward and backward procedures to generate the bootstrap series. The irreversibility of the nonlinear time series automatically excludes the latter procedures, that is, those based on the non-feasible assumption, in the nonlinear domain, that the pseudo-data are generated starting from the last p observations.

In the so-called *forward bootstrap* domain, Pan and Politis (2016) distinguish two approaches: the first considers the *fitted residuals*, whereas the second considers the *predictive residuals* that, following Politis (2013), should be favored to limit the finite-sample under-coverage of the former approach.

The steps of the *forward bootstrap* with *fitted residuals* are as follows:

PP1. the same as step L1;

PP2. compute the fitted values $\hat{\epsilon}_t = X_t - \hat{X}_t$, for $t = p+1, \ldots, T$;

PP3. center the residuals $\hat{\epsilon}_t$ and draw the bootstrap residuals, ϵ_t^*, extracting with replacement from the centered $\hat{\epsilon}_t$;

PP4. generate $T + m$ artificial data from model [15.2] using as first p pseudo-observations the vector $(X_k, X_{k+1}, X_{k+p-1})$, for $k = 1, \ldots, T-p+1$, randomly selected form the observed series such that

$$\begin{aligned} X_t^* = & \left(\hat{\phi}_0^{(1)} + \hat{\phi}_1^{(1)} X_{t-1}^* + \ldots + \hat{\phi}_p^{(1)} X_{t-p}^*\right) \mathbb{I}_{\{X_{t-1}^* \in \hat{\mathbb{R}}_1\}} \\ & + \left(\hat{\phi}_0^{(2)} + \hat{\phi}_1^{(2)} X_{t-1}^* + \ldots + \hat{\phi}_p^{(2)} X_{t-p}^*\right) (1 - \mathbb{I}_{\{X_{t-1}^* \in \hat{\mathbb{R}}_1\}}) + \epsilon_t^* \end{aligned} \quad [15.8]$$

for $t = p+1, \ldots, T$ and $\hat{\mathbb{R}}_1$ is the real subset $(-\infty, \hat{r}_1]$. Then, discard from the pseudo-series the first m artificial data;

PP5. estimate the parameters of the SETAR(2; p, p) model using the pseudo-data and then compute the bootstrap forecasts:

$$\begin{aligned} \hat{X}_{T+1}^{ce*} = & \left(\hat{\phi}_0^{(1)*} + \hat{\phi}_1^{(1)*} X_T^* + \ldots + \hat{\phi}_p^{(1)*} X_{T-p+1}^*\right) \mathbb{I}_{\{X_T^* \in \hat{\mathbb{R}}_1^*\}} \\ & + \left(\hat{\phi}_0^{(2)*} + \hat{\phi}_1^{(2)*} X_T^* + \ldots + \hat{\phi}_p^{(2)*} X_{T-p+1}^*\right) (1 - \mathbb{I}_{\{X_T^* \in \hat{\mathbb{R}}_1^*\}}) \end{aligned} \quad [15.9]$$

where $X_t^* = X_t$, for $t = T-p+1, \ldots, T$ and $\hat{\mathbb{R}}_1^*$ is the real subset $(-\infty, \hat{r}_1^*]$;

PP6. generate the future bootstrap observations

$$\begin{aligned} X_{T+1}^* = & \left(\hat{\phi}_0^{(1)} + \hat{\phi}_1^{(1)} X_T^* + \ldots + \hat{\phi}_p^{(1)} X_{T-p+1}^*\right) \mathbb{I}_{\{X_T^* \in \hat{\mathbb{R}}_1\}} \\ & + \left(\hat{\phi}_0^{(2)} + \hat{\phi}_1^{(2)} X_T^* + \ldots + \hat{\phi}_p^{(2)} X_{T-p+1}^*\right) (1 - \mathbb{I}_{\{X_T^* \in \hat{\mathbb{R}}_1\}}) + \epsilon_{T+1}^* \end{aligned} \quad [15.10]$$

with $X_t^* = X_t$, for $t = T-p+1, \ldots, T$;

PP7. calculate the bootstrap prediction error $X^*_{T+1} - \hat{X}^{ce*}_{T+1}$ (called the *bootstrap roots* in Pan and Politis (2016));

PP8. repeat B times the steps PP3–PP7 and compute the $\alpha/2$ and $1-\alpha/2$ quantile (denoted $q_{(\alpha/2)}$ and $q_{(1-\alpha/2)}$) from the empirical distribution of the bootstrap prediction error;

PP9. compute the one-step-ahead prediction from [15.5], $\hat{X}^{ce}_{T+1}$, using the parameters estimated in PP1;

PP10. construct for X_{T+1} the prediction interval:

$$[\hat{X}^{ce}_{T+1} + q_{(\alpha/2)}, \hat{X}^{ce}_{T+1} + q_{(1-\alpha/2)}]$$

The main difference of the forward bootstrap *predictive residuals*, with respect to the fitted residuals, is the following: given the time series X_t, delete the single observation X_t, for $t = p+1, \ldots, T$, from the series and then estimate the parameters $\hat{\phi}_i^{(k,t)}$, $\hat{r}^{(t)}$, for $k = 1, 2$ and $i = 1, \ldots, p$.

The fitted values described in step PP2 are now obtained using the $\hat{\phi}_i^{(k,t)}$, $\hat{r}^{(t)}$ parameters and so the *predictive residuals* becomes $\hat{\epsilon}_t^{(t)} = X_t - \hat{X}_t^{(t)}$, for $t = p+1, \ldots, T$.

In the previous algorithm, the predictive residuals replace the fitted ones starting from the step PP3; at the same time, even though their computation makes the algorithm more heavy, they make it possible to face some coverage problems that are encountered in the fitted residual case, when T is small. The algorithms Li (2011), Pan and Politis (2016) and Staszewska-Bystrova and Winker (2016) have been implemented for the predictor [15.4] where the $\hat{X}^{ce*}_{T+1}$ in the steps L3 and PP5 (and, obviously, in all subsequent steps that refer to this predictor), respectively, is replaced with the predictor $\hat{X}^{w*}_{T+1}$, whose weights are computed using the bootstrap series.

In more detail, the steps L3 and PP5 are substituted with:

L3′. estimate the model parameters using the bootstrap series X_t^* and so compute the one-step-ahead forecast $\hat{X}^{w*}_{T+1}$

$$\hat{X}^{w*}_{T+1} = \sum_{t=p+1}^{T} \hat{w}_t^* X_t^* + \epsilon^*_{T+1}$$

where the weights w_t^*, $t = p+1, \ldots, T$ are estimated from the bootstrap series.

PP5′. estimate the parameters of the SETAR(2; p, p) model using the pseudo-data and then compute the bootstrap forecasts

$$\hat{X}^{w*}_{T+1} = \sum_{t=p+1}^{T} \hat{w}_t^* X_t^*$$

where the weights w_t^*, $t = p+1, \ldots, T$ are estimated from the bootstrap series.

The four bootstrap procedures (Li (2011) and Staszewska-Bystrova and Winker (2016), the *fitted* and *predictive* approaches of Pan and Politis (2016)) described here show an increasing computational effort and a clear main difference: Li (2011) and Staszewska-Bystrova and Winker (2016) build the PI's using the percentile of the bootstrap distribution of the predictor and Pan and Politis (2016) consider the percentile of the bootstrap roots (defined in PP7) taking advantage of some results in Politis (2013) in the regression domain.

15.3. Monte Carlo simulation

To evaluate the performance of the bootstrap procedures to obtain the PI's of the weighted predictor, we have considered the approaches of Li (2011), Pan and Politis (2016) and Staszewska-Bystrova and Winker (2016) adapted to the predictor [15.4], discussed in section 15.2. We have implemented a Monte Carlo simulation study where we have considered three different SETAR(2; 1,1) models

$$M1: \; X_t = 0.20X_{t-1}\mathbb{I}_{\{X_{t-1}\in\mathbb{R}_1\}} + 0.80X_{t-1}(1 - \mathbb{I}_{\{X_{t-1}\in\mathbb{R}_1\}}) + \epsilon_t$$

$$M2: \; X_t = 0.63X_{t-1}\mathbb{I}_{\{X_{t-1}\in\mathbb{R}_1\}} - 1.3X_{t-1}(1 - \mathbb{I}_{\{X_{t-1}\mathbb{R}_1\}}) + \epsilon_t$$

$$M3: \; X_t = (-0.30 + 0.35X_{t-1})\mathbb{I}_{\{X_{t-1}\in\mathbb{R}_1\}} + (1 - 0.8X_{t-1})(1 - \mathbb{I}_{\{X_{t-1}\in\mathbb{R}_1\}}) + \epsilon_t$$

with $\mathbb{R}_1$ being the subset of not positive real numbers and $\epsilon_t \sim N(0,1)$.

Model M1 is the parametrization of the SETAR model considered in Li (2011) and Staszewska-Bystrova and Winker (2016). It is characterized by an unequal distribution of the observations between the two regimes with, on average, less than 30% of data belonging to the first regime. Model M2 is characterized by a more persistent second regime (compared to model M1), whereas model M3 has been chosen to evaluate whether the intercepts affect the forecast accuracy of the weighted predictor.

We have considered 2000 Monte Carlo replications, and in each replicate, we have considered $B = 2000$ bootstrap pseudo-series that make it possible to obtain, for the weighted predictor, the PI whose lower and upper bounds are denoted with L_i and U_i, respectively, for $i = 1, \ldots, 2000$. The empirical coverage of the interval $[L_i, U_i]$ has then been assessed, generating 2000 values from the SETAR(2; 1,1) model

$$X_{T+1,j} = \hat{\phi}_1^{(1)} X_T \mathbb{I}_{\{X_T\in\hat{\mathbb{R}}_1\}} + \hat{\phi}_1^{(2)} X_T(1 - \mathbb{I}_{\{X_T\in\hat{\mathbb{R}}_1\}}) + \epsilon_j^*$$

where X_T is the observation at time T, $(\hat{\phi}_1^{(1)}, \hat{\phi}_1^{(2)}, \hat{r}_1)$ is the vector of the estimated parameters, both obtained from the series generated at each Monte Carlo iteration. The threshold estimate, $\hat{r}_1$, is obtained defining a grid of values, delimited from the 15th and 85th percentile of X_t, such that $\sum_{t=p+1}^{n} \hat{\epsilon}_t^2$ is minimized. Finally, ϵ_j^* is the bootstrap error randomly selected (with replacement) in the jth bootstrap replicate, for $j = 1, \ldots, 2000$, of the Monte Carlo iterations.

Given this, the empirical coverage of the bootstrap prediction interval is evaluated by first computing

$$CVR_i = \frac{1}{2000} \sum_{j=1}^{2000} \mathbb{I}_{\{X_{T+1,j} \in [L_i, U_i]\}}$$

and then

$$CVR = \frac{1}{2000} \sum_{i=1}^{2000} CVR_i \qquad LEN = \frac{1}{2000} \sum_{i=1}^{2000} (U_i - L_i),$$

where LEN is the average length of the bootstrap intervals.

The CVR and LEN indices for the four bootstrap approaches adapted to the predictor [15.4] (Li (2011) and Staszewska-Bystrova and Winker (2016) and the fitted and predictive bootstrap of Pan and Politis (2016)) are reported in Table 15.1, considering time series of length $T = \{100, 200\}$ at two different nominal coverages, $1-\alpha = 0.95$ and $1-\alpha = 0.90$. The variability of the length of the 2000 PI's, obtained from the Monte Carlo iterations, is further evaluated with the standard errors (s.e.).

If we examine the results in Table 15.1, we can observe that with model M1 and $T = 100$, the PI obtained from the Li (2011) and Staszewska-Bystrova and Winker (2016) algorithm is characterized by empirical coverage (CVR) greater than the nominal PI. This over-coverage could be due to the structure of the generating process whose asymmetry in the distribution of the observations between the two regimes affects not only the weighted point predictor (as pointed out in section 15.2) but even its accuracy. On the contrary, an over-coverage that is so high is not recognized in the PI obtained from the *fitted* and *predictive* approaches of Pan and Politis (2016) algorithms and even in all cases with $T = 200$, with the exception of the Staszewska-Bystrova and Winker (2016) algorithm.

The results of models M2 and M3 show the main differences when $T = 100$. In this case, the empirical coverage of the fitted bootstrap outperforms the other approaches, whereas the predictive bootstrap approach is always characterized by PI's with the highest width and even the highest variability of their length. It can be further noted that all bootstrap approaches benefit from the presence of the intercepts in the SETAR model, as in the M3 case, where the length of the PI's and its variability are smaller than in the M1 and M2 models. This has different explanations: first, the presence of the intercepts makes it possible to better discriminate between the two regimes and so even the estimation of the SETAR model takes advantage of it; second, as said before, the weighted predictor has better performance when the discrimination among regimes is more marked.

	Nominal coverage 0.95						Nominal coverage 0.90					
	$T = 100$			$T = 200$			$T = 100$			$T = 200$		
	CVR	LEN	s.e.	CVR	LEN	s.e.	CVR	LEN	s.e.	CVR	LEN	s.e.
Model M1												
w_fit	0.9577	6.2094	0.9537	0.9494	5.4729	0.4682	0.9066	5.0229	0.6352	0.8992	4.5825	0.3799
w_pred	0.9504	6.4031	1.4113	0.9501	5.4965	0.4715	0.8991	5.1296	0.9018	0.8995	4.6046	0.3848
w_Li	0.9695	6.1093	1.2617	0.9556	5.3762	0.5374	0.9245	4.9035	0.6071	0.9078	4.5012	0.3825
w_SW	0.9750	9.0207	1.2025	0.9605	5.4734	0.5340	0.9368	5.0793	0.8919	0.9162	4.5911	0.3603
Model M2												
w_fit	0.9465	6.9169	0.9909	0.9491	4.7688	0.2998	0.8985	5.1854	0.5569	0.8997	3.9444	0.2325
w_pred	0.9602	7.2943	1.3652	0.9520	4.8868	0.3102	0.9161	5.4593	0.7510	0.9026	4.0406	0.2430
w_Li	0.9528	6.8631	1.0304	0.9478	4.7157	0.2948	0.9049	5.1132	0.5481	0.8982	3.9003	0.2323
w_SW	0.9563	7.0630	0.9491	0.9481	4.8272	0.1823	0.9127	5.2970	0.4680	0.9083	3.9934	0.1329
Model M3												
w_fit	0.9509	5.1458	0.4444	0.9495	4.6117	0.2509	0.9010	4.2817	0.3458	0.9000	3.8756	0.2069
w_pred	0.9580	5.3809	0.5834	0.9507	4.7108	0.2595	0.9110	4.4637	0.4265	0.9014	3.9489	0.2161
w_Li	0.9344	5.0661	0.4633	0.9389	4.5630	0.2558	0.8829	4.2109	0.3251	0.8901	3.8332	0.2123
w_SW	0.9438	5.2547	0.4105	0.9456	4.6650	0.2245	0.8975	4.3703	0.3238	0.8998	3.9094	0.2078

Table 15.1. *Evaluation of the PI's of the weighted predictor for the models M1, M2 and M3, at two nominal coverages, 0.95 and 0.90, and two different series lengths,* $T = \{100, 200\}$. w_fit *and* w_pred*: fitted bootstrap and predictive bootstrap PI's based on Pan and Politis (2016);* w_Li*: bootstrap PI based on Li (2011);* w_SW *bootstrap PI based on Staszewska-Bystrova and Winker (2016)*

	Nominal coverage 0.95			Nominal coverage 0.90		
	CVR	LEN	s.e.	CVR	LEN	s.e.
w_Li	0.9387	5.0600	0.3318	0.8901	4.2091	0.2442
w_SW	0.9451	5.2374	0.4395	0.8981	4.3614	0.3398

Table 15.2. *Skewness-adjusted (Grabowski* et al. *2020) PI's of Li (2011) and Staszewska-Bystrova and Winker (2016) of the weighted predictor [15.4] for model M3, at two nominal coverages, 0.95 and 0.90 and series length* $T = 100$

The results obtained from model M3 have been further investigated. The deviation from the empirical and the nominal coverage of the PI's (mainly of the Li (2011) and Staszewska-Bystrova and Winker (2016) algorithms) has led to construct skewness-adjusted confidence intervals. In particular, following Grabowski *et al.* (2020), we mirror the bootstrap distribution of the weighted predictor [15.4] using the mirrored bootstrap prediction

$$\hat{X}^{w*m}_{T+1} = \hat{X}^{w}_{T+1} - (\hat{X}^{w*}_{T+1} - \hat{X}^{w}_{T+1})$$

in the steps L3' and PP5'.

The results of model M3 with T = 100 are shown in Table 15.2, where it can be noted that the coverage of the PI's, the length and its variability do not take much advantage of the correction of Grabowski *et al.* (2020) in the presence of one-step-ahead forecasts. It is in line with their results but we expect that their approach will give more clear results when the number of steps ahead increases.

From the computational point of view, the burden of the four algorithms is quite heavy. All simulations have been carried out with a processor Intel Core i7 quad core, 3.3 GHz and, on average, the computing time (in seconds) of each Monte Carlo iteration for the four bootstrap methods, with series length $T = 100$ are: fitted bootstrap 33.51", predictive bootstrap 34.45", the Li (2011) method 29.68" and the Staszewska-Bystrova and Winker (2016) method 29.60". As expected, the last two approaches are less heavy from the computational point of view, while the best coverage of the Pan and Politis (2016) algorithm is related to a longer computation time.

Finally, we need to remark that the results given here can be further expanded evaluating how the distribution of the errors can impact the predictive accuracy (and the PI's coverage) of the predictor [15.4]. This point and further evaluations on the bootstrap PI's are left for the future research.

15.4. References

Boero, G. and Marrocu, M. (2004). The performance of SETAR models: A regime conditional evaluation of point, interval and density forecasts. *International Journal of Forecasting*, 20, 305–320.

Bühlmann, P. (2002). Bootstraps for time series. *Statistical Science*, 17, 52–72.

Clements, M.P. and Smith, J. (1997). The performance of alternative forecasting methods for SETAR models. *International Journal of Forecasting*, 13, 463–475.

Clements, M.P., Franses, J.S., van Dijk, D. (2003). On SETAR non-linearity and forecasting. *Journal of Forecasting*, 22, 359–375.

Grabowski, D., Staszewska-Bystrova, A., Winker, P. (2020). Skewness-adjusted bootstrap confidence intervals and confidence bands for impulse response functions. *AStA Advances in Statistical Analysis*, 104, 5–32.

Kreiss, J.P. and Paparoditis, E. (2011). Bootstrap methods for dependent data: A review. *Journal of Korean Statistical Society*, 40, 357–378.

Li, J. (2011). Bootstrap prediction intervals for SETAR models. *International Journal of Forecasting*, 27, 320–332.

Niglio, M. (2019). SETAR forecasts with weighted observations. *39th International Symposium on Forecasting*, Thessaloniki, Greece.

Pan, L. and Politis, D.N. (2016). Bootstrap prediction intervals for linear, nonlinear and nonparametric autoregressions. *Journal of Statistical Planning and Inference*, 177, 1–27.

Politis, D. (2013). Model-free model-fitting and predictive distributions (with discussions). *Test*, 22, 183–250.

Staszewska-Bystrova, A. and Winker, P. (2016). Improved bootstrap prediction intervals for SETAR models. *Statistical Papers*, 57, 89–98.

Tong, H. (1978). On a threshold model. In *Pattern Recognition and Signal Processing*, Chen, C.H. (ed.). Sijthoff and Noordhoff, Amsterdam.

Tong, H. (1983). *Threshold Models in Non-linear Time Series Analysis*. Springer-Verlag, New York.

Tong, H. (1990). *Non-linear Time Series: A Dynamical System Approach*. Oxford University Press, New York.

16

Revisiting Transitions Between Superstatistics

This work aims to provide an accurate method for the detection of a transition between superstatistics. A slight improvement over the currently published method is achieved. The superstatistics framework is briefly recalled and a rather new concept of the transition of superstatistics, introduced by Xu and Beck (2016), is re-examined. In addition, an original synthetic model for superstatistical transition, suggested by Beck, is discussed. It is shown that its modified version, which takes into account a stochastic nature of the transition, better reflects empirically observed transitions.

16.1. Introduction

Superstatistics is a well-known term in the field of non-equilibrium statistical physics. It describes a system in a local thermodynamic equilibrium. However, only recently, a new spin in the superstatistical paradigm has been introduced, namely, a transition of superstatistics (Xu and Beck 2016). Its application is predominantly in time series analysis. The basic premise is that the superstatistical smearing distribution may change on different time scales. The first experimental evidence for this phenomenon was introduced by Xu and Beck (2016). The pioneering paper was followed by our paper (Jizba *et al.* 2018) which introduced a more reliable method for detecting the alleged transition. The method was based on leveraging statistical distances in order to decide a favorable probability distribution on various time scales. Nevertheless, there were doubts about a significance level. Therefore, in this chapter we will revisit the procedure and use the Monte Carlo method to provide the probability of the successful determination of a correct smearing distribution. By doing so, we can assess a significance level to the transition.

Chapter written by Petr JIZBA and Martin PROKŠ.

16.2. From superstatistic to transition between superstatistics

Superstatistics is a concept devised by Beck for systems with fluctuating intensive parameters, for example temperature, which are therefore in a non-equilibrium state. The idea first appeared in the paper (Beck 2001), and the term superstatistics was coined later in the successive paper (Beck and Cohen 2003).

The assumption is that the system is in a non-equilibrium steady state and is composed of many cells that are locally in equilibrium, but with different values of the intensive parameter, for example temperature. This intensive parameter in each cell changes on a long time scale, much larger than the relaxation time of the cell.

Superstatistics is the generalized Boltzmann factor that describes the whole system composed of small subsystems in local equilibrium

$$B(E) = \int_0^{+\infty} f(\beta)e^{-\beta E}\mathrm{d}\beta, \qquad [16.1]$$

where $f(\beta)$ is the smearing distribution.

In general, the only restriction on $f(\beta)$ is allowing only positive values for β. However, experience showed that three distributions fit various empirical data especially well. Therefore, Beck (2009) suggested the so-called universality classes. It contains a *gamma distribution*

$$f(\beta) = \frac{1}{\Gamma(\frac{n}{2})}\left(\frac{n}{2\beta_0}\right)^{\frac{n}{2}} \beta^{\frac{n}{2}-1}\exp\left(-\frac{n\beta}{2\beta_0}\right), \qquad [16.2]$$

a *log-normal distribution*

$$f(\beta) = \frac{1}{\beta\sqrt{2\pi s^2}}\exp\left(-\frac{(\log\beta - \log\mu)^2}{2s^2}\right) \qquad [16.3]$$

and an *Inverse* χ^2 *distribution*. The last distribution is disregarded here because it was shown in Jizba *et al.* (2018) that it gives poor results for data at our disposal.

Superstatistics is definitely a great idea which has justifiable motivation; furthermore, it has been successfully fitted on empirical data. Therefore, there is little doubt about the usefulness of superstatistics; however, a new broader model was recently introduced by the father of superstatistics in Xu and Beck (2016). It is claimed that the transition of superstatistics is possible when we look at a time series at different time scales.

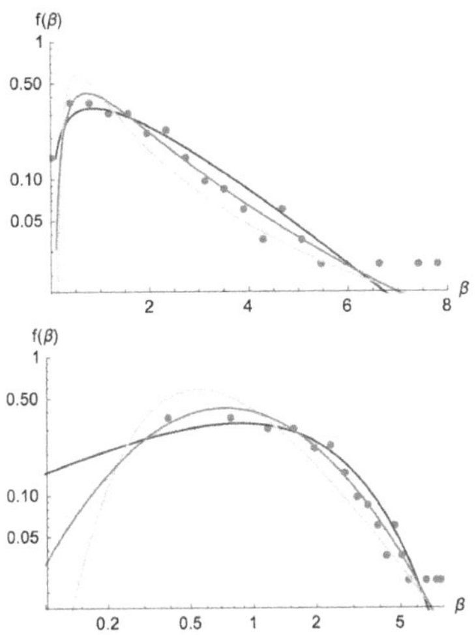

Figure 16.1. *Visual histogram comparison is difficult. For a color version of this figure, see www.iste.co.uk/dimotikalis/analysis2.zip*

The transition between superstatistics means a change from one smearing distribution to another. For example, in a financial time series, log-returns may be well described by the log-normal distribution at a small (minute) time scale; however, if one moves to the higher time scale (hours), a better description may be achieved by using a gamma distribution. This kind of superstatistical transition is examined in the next section.

16.3. Transition confirmation

Originally in the pioneering paper (Xu and Beck 2016), the transition was only assessed by looking at the histogram of β at two time scales (minutes and days). Unfortunately, it is impossible to reliably detect the transition merely from the histogram, see Figure 16.1. Therefore, in Jizba *et al.* (2018), a method based on statistical distances was employed. It allowed us to see a change from one superstatistic to another in a quantitative way. However, this still lacked a level of significance because the fact that statistical distance is smaller for one distribution than for another, may just be a manifestation of a random error. A slight improvement described in the following attempts to address this issue.

The main difference from the method in Jizba *et al.* (2018) is that we try to assign a probability distribution to each time scale according to all three statistical distances. Kolmogorov–Smirnov distance

$$D_n = \sup_x |F_n(x) - F(x)|, \tag{16.4}$$

Cramér–von Mises distance

$$C_n = n \int_{-\infty}^{+\infty} (F_n(x) - F(x))^2 \mathrm{d}F(x) \qquad [16.5]$$

and Anderson–Darling distance

$$A_n = n \int_{-\infty}^{+\infty} \frac{(F_n(x) - F(x))^2}{F(x)(1 - F(x))} \mathrm{d}F(x), \qquad [16.6]$$

where $F(x)$ is the fully specified distribution function and $F_n(x)$ is the empirical distribution function

$$F_n(x) = \frac{1}{n} \sum_{i=1}^{n} I(u_i \leq x). \qquad [16.7]$$

A correct distribution at a given time scale is considered a distribution that is favorable by at least two distances (a majority vote). Then, the ideal procedure would be to obtain the probability of successfully discriminating between the two distributions for each statistical distance, hence assigning a value of significance for the chosen distribution at each time scale. This would allow us to claim that at a time scale, for example 20 $minutes$, the smearing distribution is probably better described by a log-normal distribution, while at time scale 300 $minutes$, it is very likely a gamma distribution. Hence, the transition may be considered as significant.

Unfortunately, the statistical properties of the distances cannot be obtained if parameters are estimated, the only possibility is to use Monte Carlo simulations to determine the probability of successfully detected transition. The same procedure is used in the Lilifors test that uses Kolmogorov–Smirnov distance (see Lilliefors (1967)).

Apart from the Lilliefors test, the method of recognizing probability distribution was inspired by Marshall *et al.* (2001), where only the Kolmogorov–Smirnov distance was used to discriminate between two-parametric distribution families. It was shown that the distance measure provides a reliable discriminating criteria.

The Monte Carlo simulation is conducted as follows:

1) Estimate parameters for gamma and log-normal distribution from data by using the Maximum-Likelihood method.

2) Generate a random sample, of the size available for a particular time scale and company, from gamma or log-normal distribution (using parameters from the previous step) depending on a distribution favorable by probability distances.

3) For both gamma and log-normal distributions, calculate probability distances from the empirical distribution (estimated from the generated sample).

4) Use the decision criteria for choosing gamma or log-normal distribution and mark the trial as successful if the distribution matches the one generated in step 2.

5) Repeat steps 2 – 4 10^5 times and estimate the probability of successfully selecting the probability distribution by relative frequencies.

The dataset used for testing this method is the same as the one used in Xu and Beck (2016), i.e. stock prices of seven US companies from different sectors recorded on the minute-tick basis during a period from January 2, 1998, to May 22, 2013. The output of the simulation is depicted in Figure 16.5. It confirms the conclusion from Jizba *et al.* (2018) about a transition for the companies Alcola Inc. (AA) and Wal-Mart Stores Inc. (WMT). Moreover, it is seen that the time series for Bank of America does not exhibit a transition of superstatistics. The key point to note is a relatively high probability of successfully discriminating the two distributions. Therefore, it may be concluded that the transition, especially for AA, is not smooth, but oscillates between the two distributions around the transition point. This statistically significant observation is examined in the next section.

16.4. Beck's transition model

In the original mention of superstatistical transition (Xu and Beck 2016), Beck and Xu suggested the so-called *synthetic model*

$$\beta_\tau = \kappa_\tau L_{\tau_0} + \big(1 - \kappa_\tau\big) G_{\tau_\infty}, \tag{16.8}$$

where L_{τ_0} and G_{τ_∞} are the two random variables with log-normal and gamma distributions, respectively. The suffixes τ_0 and τ_∞ denote small and large time scales, respectively, where the distribution of β is log-normal and gamma. For data at hand, $\tau_0 = 20\ minutes$ and $\tau_\infty \approx 500\ minutes$. L_{τ_0} and G_{τ_∞} may be thought of as asymptotic distributions. The parameter $\kappa \in \langle 0, 1\rangle$ is a function of a time scale τ and is responsible for a smooth transition from a region dominated by the log-normal distribution to one with gamma distribution on larger time scales. A reasonable functional form for κ which may reproduce the observed transition is

$$\kappa(\tau) = \frac{1}{2}\Big(tanh\big(a(\tau - b)\big) + 1\Big). \tag{16.9}$$

The parameter a controls the sharpness of a transition and b selects a time scale at which the transition occurs. See Figure 16.2 for demonstration of the sharpness parameter.

However, as Monte Carlo simulations show, a deterministic increasing function of a time scale does not reflect the observed rough evolution of the transition. (Compare Figure 16.3, where a transition generated from the synthetic model equation [16.8] with κ given by equation [16.9] is depicted, and Figure 16.5, which shows the observed transitions for the companies Alcola Inc. (AA) and Wal-Mart Stores Inc. (WMT)).

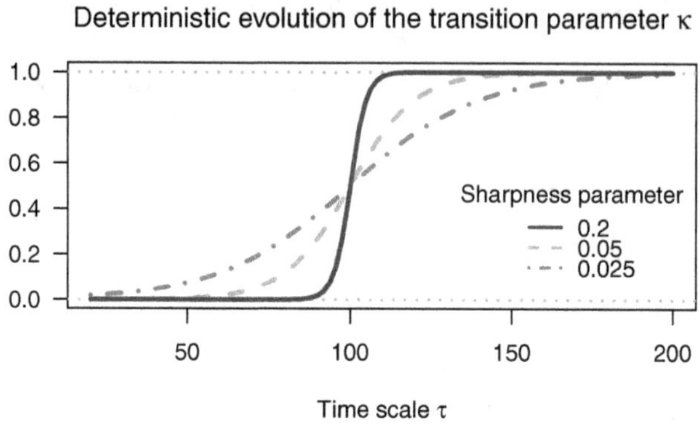

Figure 16.2. *Deterministic evolution of the transition parameter. For a color version of this figure, see www.iste.co.uk/dimotikalis/analysis2.zip*

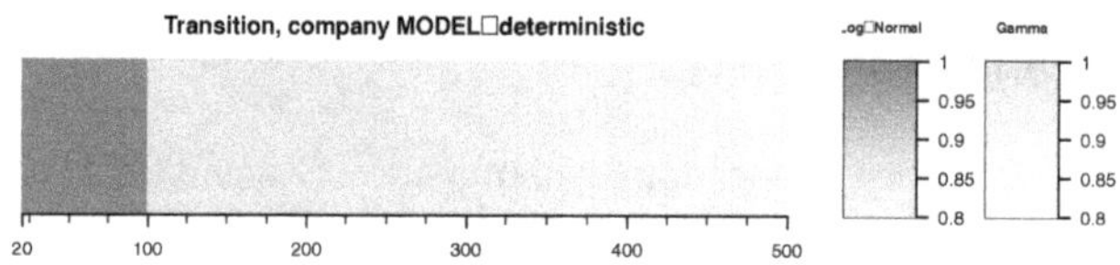

Figure 16.3. *Transition for the deterministic model. For a color version of this figure, see www.iste.co.uk/dimotikalis/analysis2.zip*

The explanation for this discrepancy is rather simple. Since the distance measures between probability distributions used for discriminating the two regions have a significant statistical power (due to a large sample size especially at small time scales), it will, at a certain level of κ (likely $\kappa \approx \frac{1}{2}$), flip from the log-normal distribution to the gamma distribution and will never oscillate between those two states (as seen in Figure 16.3).

In this chapter, we propose a model that better captures the observed behavior of real transitions. As can be seen in Figure 16.5, the transitions possess a stochastic nature. For example, time series for Wal-Mart Stores Inc. shows the transition from log-normal to the gamma region around a time scale of 60 $minutes$. Nevertheless, a quick unpredictable transition happens much sooner and also on higher time scale, where an occasional flip to the log-normal region is observed. The stochastic nature is more pronounced for Alcola Inc., where the transition again occurs around a time scale of 60 $minutes$, but unlike for WMT, it is very slow (corresponds to a small sharpness parameter in [16.9]). Even at $\tau \geq 300$ $minutes$, an occasional flip back to the log-normal distribution is observed.

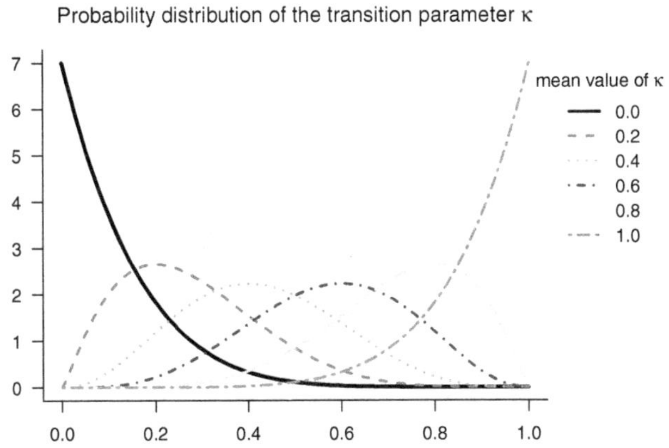

Figure 16.4. *Probability distribution of the transition parameter. For a color version of this figure, see www.iste.co.uk/dimotikalis/analysis2.zip*

The suggested modification, which incorporates a random element into the model, is to consider κ as a random variable. (Strictly speaking, a stochastic process since κ is parametrized by a time scale τ). κ is a parameter in $\langle 0, 1\rangle$; therefore, it is necessary to use a probability distribution with a compact support. A well-known distribution with this property is a beta distribution

$$p(x) = \frac{x^{\gamma-1}(1-x)^{\delta-1}}{B(\gamma,\delta)}, \qquad \gamma, \delta > 0. \qquad [16.10]$$

The original parametrization is not the best choice, therefore, an alternative one which contains the mode of the distribution μ and the so-called concentration ν is used

$$\gamma = \mu(\nu - 2) + 1,$$
$$\delta = (1 - \mu)(\nu - 2) + 1.$$

The concentration, together with sharpness parameter in [16.9], controls how centered the transition is and should complement each other, i.e. they should act as a one degree of freedom. The mode parameter μ is then time scale dependent and its functional form is given by equation [16.9]. Figure 16.4 shows the shape of the probability distribution for various time scales. For $\tau \approx \tau_0$, there is a low probability for κ to leave a log-normal region, while for $\tau \approx \tau_\infty$, κ predominantly stays in a gamma region.

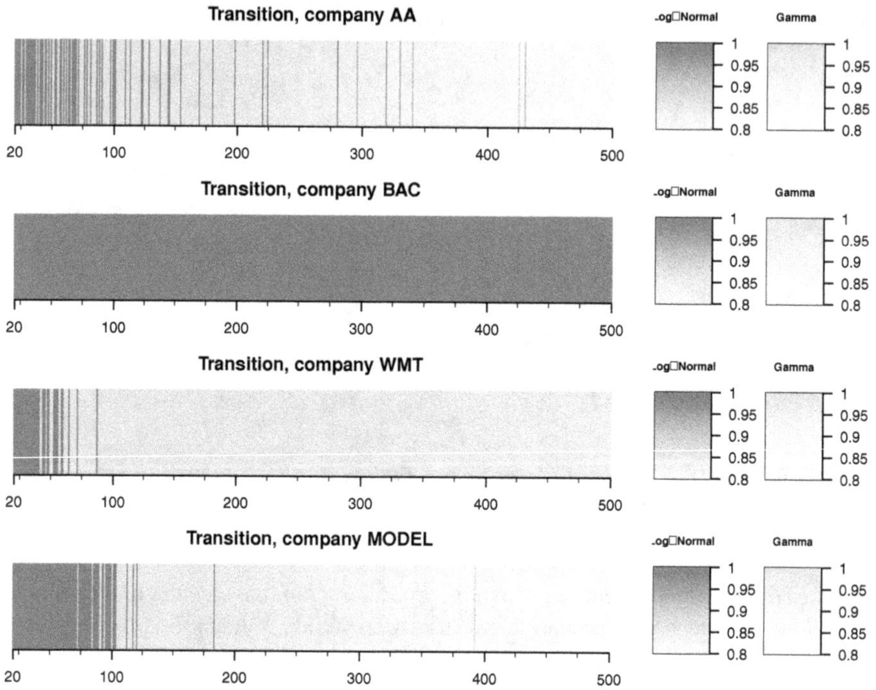

Figure 16.5. *Transitions for companies Alcola Inc. (AA), Bank of America Corporation (BAC), Wal-Mart Stores Inc. (WMT) and the synthetic model incorporating randomness. For a color version of this figure, see www.iste.co.uk/dimotikalis/analysis2.zip*

It should be noted that even though the model reflects the empirical transition well, a suitable estimator for corresponding parameters in the model has not been found.

16.5. Conclusion

The Beck's synthetic model for superstatistical transition was revisited. It was shown that its modified version, which involves a random element, is able to correctly reproduce observed transitions and may therefore serve as a suitable model. The modification is done by incorporating a random element into the transition parameter κ. Namely, κ is considered to be a stochastic process (parametrized by a time scale) with beta distribution. Moreover, a better method for assessing the transition of superstatistics was provided, which assigns the probability of successfully discriminating between two Superstatistical regions. These probabilities need to be obtained by Monte Carlo simulations.

16.6. Acknowledgments

This work was supported by the Grant Agency of the Czech Technical University in Prague (grant no. SGS19/239/OHK4-009/19).

16.7. References

Beck, C. (2001). Dynamical foundatiouns of nonextensive statistical mechanics. *Phys. Rev. Lett.*, 87.

Beck, C. (2009). Recent developments in superstatistics. *Braz. J. Phys.*, 39(2A).

Beck, C. and Cohen, E. (2003). Superstatistics. *Physica A*, 322, 267.

Jizba, P., Korbel, J., Lavička, H., Prokš, M., Svoboda, V., Beck, C. (2018). Transitions between superstatistical regimes: Validity, breakdown and applications. *Physica A*, 493, 29–46.

Lilliefors, H. (1967). On the Kolmogorov–Smirnov test for normality with mean and variance unknown. *J. Am. Stat. Assoc.*, 62, 399–402.

Marshall, A.W., Meza, J., Olkin, I. (2001). Can data recognize its parent distribution? *J. Comput. Graph. Stat.*, 10, 555–580.

Xu, D. and Beck, C. (2016). Transition from lognormal to superstatistics for financial time series. *Physica A*, 453, 173–183.

17

Research on Retrial Queue with Two-Way Communication in a Diffusion Environment

In this chapter, we consider a retrial queuing system in which incoming fresh calls arrive at the server according to a Poisson process. Upon arrival, an incoming call either occupies the server if it is idle, or joins an orbit if the server is busy. From the orbit, an incoming call retries to occupy the server and behaves as a fresh incoming call. After some idle time, the server makes an outgoing call to the outside. The system operates in a random environment. Random external factors affect the service time of applications. A random medium is represented by a diffusion process. For this system, we obtained the probability distribution of the states of the server and the probability distribution of a number of calls in the system.

17.1. Introduction

Modern service systems, including call centers, strive to minimize downtime. Minimizing downtime increases the efficiency of the system. For the promotion of call center services, operators can make outgoing calls. The following chapter deals with the two-way communication queue (Artalejo and Gomez-Corral 2008).

In addition, the work of operators depends on a number of random factors. Such factors directly affect the success of the call and the service time. Generally, such factors are called a random environment. Information on the functioning of such systems is of great practical interest. Purdue first introduced the concept of a queuing system in a random environment (Purdue 1974). The first work to study the characteristics of a queuing system in a Markovian random environment with two

---

Chapter written by Viacheslav VAVILOV.

states was carried out by Yechiali and Naor (1971). This result was soon generalized by Yechiali (1973), who used the Markov chain with an arbitrary finite number of states as the controlling process. Purdue obtained the condition for the existence of a stationary mode in such a queuing system.

Neuts (1971, 1978) reduced the task of studying this queuing system in a random environment for a particular case, to solving a matrix equation.

There are models in which the random environment affects not only the operation of the device, but also the parameters of the input request stream (Nazarov and Phung-Duc 2019).

The problem of developing new research methods and modifying available research methods for outgoing flows is quite relevant.

In this chapter, the main method of research is the asymptotic analysis method, which makes it possible to find the main probabilistic characteristics of the system in the asymptotic condition of a large delay in orbit. The most important characteristic is the probability distribution density of device states.

17.2. Mathematical model

Consider a retrial queue with an incoming flow of Poisson with the parameter λ. An application received for a free device begins to be serviced. Service time is exponential with parameter μ_1. The submitted application leaves the system. An attempt to receive a new application for an occupied device leads to the transfer of an application to an orbit. Requests are resubmitted to the device after an accidental delay. The delay time has an exponential distribution with the parameter γ. The number of applications in orbit is i. An unoccupied device causes additional claims from the outer orbit with intensity θ. The service time of such an application has an exponential distribution with the parameter μ_2.

We denote the state of the device as follows: $k=0$ if it is free, $k=1$ if it is busy servicing the application and $k=2$ if the device is servicing the called application.

The system operates in a random environment. The mathematical model of a random environment is a diffusion process. The diffusion process is defined by the equation $ds(t)=\alpha(s)dt+\beta(s)dw(t)$, where $w(t)$ is the Wiener process. The random environment affects the functioning of the system, such as $\mu_1=\mu_1(s)$,

$\mu_2 = \mu_2(s)$. The probabilities of servicing requests for time Δt are equal to $\mu_1(s)\Delta t + o(\Delta t)$ and $\mu_2(s)\Delta t + o(\Delta t)$.

A random process $\{k(t), i(t), s(t)\}$ is a continuous-time Markov chain.

Let $P(k(t) = k, i(t) = i, s \leq s(t) < s + ds) / ds = P_k(i,s,t)$.

The following condition must be met:

$$\sum_{k=0}^{2}\sum_{i=0}^{\infty}\int_{-\infty}^{\infty} P_k(i,s,t)ds = 1$$

The probability distribution $P_k(i,s,t)$ satisfies the Kolmogorov system:

$$\frac{\partial P_0(i,s,t)}{\partial t} + (\lambda + i\gamma + \theta)P_0(i,s,t) = \mu_1(s)P_1(i,s,t) +$$

$$+\mu_2(s)P_2(i,s,t) - \frac{\partial}{\partial s}\{\alpha(s)P_0(i,s,t)\} + \frac{1}{2}\frac{\partial^2}{\partial s^2}\{\beta^2(s)P_0(i,s,t)\}$$

$$\frac{\partial P_1(i,s,t)}{\partial t} + (\lambda + \mu_1(s))P_1(i,s,t) = \lambda P_0(i,s,t) +$$

$$+(i+1)\gamma P_0(i+1,s,t) + \lambda P_1(i-1,s,t) -$$

$$-\frac{\partial}{\partial s}\{\alpha(s)P_1(i,s,t)\} + \frac{1}{2}\frac{\partial^2}{\partial s^2}\{\beta^2(s)P_1(i,s,t)\},$$

$$\frac{\partial P_2(i,s,t)}{\partial t} + (\lambda + \mu_2(s))P_2(i,s,t) = \theta P_0(i,s,t) +$$

$$+\lambda P_2(i-1,s,t) - \frac{\partial}{\partial s}\{\alpha(s)P_2(i,s,t)\} + \frac{1}{2}\frac{\partial^2}{\partial s^2}\{\beta^2(s)P_2(i,s,t)\}.$$

This system is investigated using the asymptotic analysis method. The asymptotic condition $\gamma \to 0$.

Let

$$\gamma=\varepsilon^2,\ \varepsilon^2 t=\tau,\ \varepsilon^2 i=x+\varepsilon y,\ \frac{1}{\varepsilon}P_k(i,s,t)=H_k(y,s,\tau,\varepsilon).$$

We obtain the system

$$\varepsilon^2\frac{\partial H_0(y,s,\tau,\varepsilon)}{\partial\tau}-\varepsilon x'(\tau)\frac{\partial H_0(y,s,\tau,\varepsilon)}{\partial y}+(\theta+\lambda+x+\varepsilon y)H_0(y,s,\tau,\varepsilon)=$$

$$=\mu_1(s)H_1(y,s,\tau,\varepsilon)+\mu_2(s)H_2(y,s,\tau,\varepsilon)-$$

$$-\frac{\partial}{\partial s}\{\alpha(s)H_0(y,s,\tau,\varepsilon)\}+\frac{1}{2}\frac{\partial^2}{\partial s^2}\{\beta^2(s)H_0(y,s,\tau,\varepsilon)\},$$

$$\varepsilon^2\frac{\partial H_1(y,s,\tau,\varepsilon)}{\partial\tau}-\varepsilon x'(\tau)\frac{\partial H_1(y,s,\tau,\varepsilon)}{\partial y}+(\lambda+\mu_1(s))H_1(y,s,\tau,\varepsilon)=$$

$$=\lambda H_1(y-\varepsilon,s,\tau,\varepsilon)+(x+\varepsilon(y+\varepsilon))H_0(y+\varepsilon,s,\tau,\varepsilon)+$$

$$+\lambda H_0(y,s,\tau,\varepsilon)-\frac{\partial}{\partial s}\{\alpha(s)H_1(y,s,\tau,\varepsilon)\}+\frac{1}{2}\frac{\partial^2}{\partial s^2}\{\beta^2(s)H_1(y,s,\tau,\varepsilon)\},$$

$$\varepsilon^2\frac{\partial H_2(y,s,\tau,\varepsilon)}{\partial\tau}-\varepsilon x'(\tau)\frac{\partial H_2(y,s,\tau,\varepsilon)}{\partial y}+(\lambda+\mu_2(s))H_2(y,s,\tau,\varepsilon)=$$

$$=\theta H_0(y,s,\tau,\varepsilon)+\lambda H_2(y-\varepsilon,s,\tau,\varepsilon)-$$

$$-\frac{\partial}{\partial s}\{\alpha(s)H_2(y,s,\tau,\varepsilon)\}+\frac{1}{2}\frac{\partial^2}{\partial s^2}\{\beta^2(s)H_2(y,s,\tau,\varepsilon)\} \quad [17.1]$$

Further research is carried out with this system.

17.3. Asymptotic average characteristics

Asymptotic average characteristics are the probability distribution $R_k(x)$ of device states and the function $x=x(\tau)$. The limiting process $x(\tau)=\lim_{\varepsilon\to 0}\left(\varepsilon^2 i(\tau/\varepsilon^2)\right)$ is the asymptotic average of the normalized number of applications in the system. We will show that it is a deterministic function.

In system [17.1], we set the limit

$$\lim_{\varepsilon\to 0} H_k(y,s,\tau,\varepsilon) = H_k(y,s,\tau),$$

and obtain the system

$$(\theta+\lambda+x)H_0(y,s,\tau) = \mu_1(s)H_1(y,s,\tau)+ \mu_1(s)H_2(y,s,\tau)-$$

$$-\frac{\partial}{\partial s}\{\alpha(s)H_0(y,s,\tau)\}+\frac{1}{2}\frac{\partial^2}{\partial s^2}\{\beta^2(s)H_0(y,s,\tau)\},$$

$$\mu_1(s)H_1(y,s,\tau) = (\lambda+x)H_0(y,s,\tau)-$$

$$-\frac{\partial}{\partial s}\{\alpha(s)H_1(y,s,\tau)\}+\frac{1}{2}\frac{\partial^2}{\partial s^2}\{\beta^2(s)H_1(y,s,\tau)\},$$

$$\mu_2(s)H_2(y,s,\tau) = \theta H_0(y,s,\tau)-$$

$$-\frac{\partial}{\partial s}\{\alpha(s)H_2(y,s,\tau)\}+\frac{1}{2}\frac{\partial^2}{\partial s^2}\{\beta^2(s)H_2(y,s,\tau)\} \quad [17.2]$$

The solution $H_k(y,s,\tau)$ of system [17.2] can be written in the following form:

$$H_k(y,s,\tau) = Q_k(x,s)H(y,\tau) \quad [17.3]$$

The function $H(y,\tau)$ is a probability density of process values. The function $Q_k(x,s)$ is a two-dimensional probability distribution of instrument states k and states s of a random environment. The function Q(x,s) is determined by the system.

$$(\theta+\lambda+x)Q_0(x,s) = \mu_1(s)Q_1(x,s)+\mu_2(s)Q_2(x,s)-$$

$$-\frac{\partial}{\partial s}\{\alpha(s)Q_0(x,s)\}+\frac{1}{2}\frac{\partial^2}{\partial s^2}\{\beta^2(s)Q_0(x,s)\},$$

$$\mu_1(s)Q_1(x,s) = (\lambda+x)Q_0(x,s)-\frac{\partial}{\partial s}\{\alpha(s)Q_1(x,s)\}+\frac{1}{2}\frac{\partial^2}{\partial s^2}\{\beta^2(s)Q_1(x,s)\},$$

$$\mu_2(s)Q_2(x,s) = \theta Q_0(x,s) - \frac{\partial}{\partial s}\{\alpha(s)Q_2(x,s)\} + \frac{1}{2}\frac{\partial^2}{\partial s^2}\{\beta^2(s)Q_2(x,s)\} \quad [17.4]$$

and the condition

$$\sum_{k=0}^{2}\int_{-\infty}^{+\infty} Q_k(x,s)ds = 1 \quad [17.5]$$

We denote

$$\sum_{k=0}^{2} Q_k(x,s) = r(s), \; \int_{-\infty}^{+\infty} Q_k(x,s)ds = R_k(x) \quad [17.6]$$

The function $R_k(x)$ is a probability distribution of device states. The function $r(s)$ is a probability distribution of a random environment state.

The following conditions must be met:

$$\int_{-\infty}^{+\infty} r(s)ds = 1, \; \sum_{k=0}^{2} R_k(x) = 1 \quad [17.7]$$

We sum the equations of system [17.4] by k and take the notation [17.5] into account, in order to obtain the following equation:

$$-\frac{\partial}{\partial s}\{\alpha(s)r(s)\} + \frac{1}{2}\frac{\partial^2}{\partial s^2}\{\beta^2(s)r(s)\} = 0 \quad [17.8]$$

This equation determines the stationary probability distribution of the states of the diffusion environment.

Equation [17.8] is a linear homogeneous differential equation, which has the solution

$$r(s) = \frac{1}{\beta^2(s)} e^{2\int_{-\infty}^{s} \frac{\alpha(u)}{\beta^2(u)}du} \Bigg/ \int_{-\infty}^{+\infty} \frac{1}{\beta^2(s)} e^{2\int_{-\infty}^{s} \frac{\alpha(u)}{\beta^2(u)}du} \quad [17.9]$$

By integrating the equations of system [17.4] by s and taking [17.6] into account, we obtain

$$\int_{-\infty}^{+\infty} \mu_1(s)Q_1(x,s)ds = \psi R_1(x)\,,\ \int_{-\infty}^{+\infty} \mu_2(s)Q_2(x,s)ds = \phi R_2(x) \qquad [17.10]$$

and make an assumption

$$\left(-\alpha(s)Q_k(x,s)+\frac{1}{2}\frac{\partial}{\partial s}\left\{\beta^2(s)Q_k(x,s)\right\}\right)\Bigg|_{s=-\infty}^{+\infty}=0,$$

system [17.4] takes the form

$$(\theta+\lambda+x)R_0(x)=\psi R_1(x)+\phi R_2(x),$$

$$\psi R_1(x)=(\lambda+x)R_0(x),$$

$$\phi R_2(x)=\theta R_0(x) \qquad [17.11]$$

System [17.11] and condition [17.7] gives the solution

$$R_0(x)=\frac{\psi\phi}{\theta\psi+(\lambda+x)\phi+\psi\phi},\quad R_1(x)=\frac{(\lambda+x)\phi}{\theta\psi+(\lambda+x)\phi+\psi\phi},$$

$$R_2(x)=\frac{\alpha\psi}{\theta\psi+(\lambda+x)\phi+\psi\phi} \qquad [17.12]$$

In system [17.1], by expanding the functions $H_k(y\pm\varepsilon,s,\tau,\varepsilon)$ in a series in increments of the argument y up to $o(\varepsilon)$, we obtain

$$-\varepsilon x'(\tau)\frac{\partial H_0(y,s,\tau,\varepsilon)}{\partial y}+(\theta+\lambda+x+\varepsilon y)H_0(y,s,\tau,\varepsilon)=$$

$$=\mu_1(s)H_1(y,s,\tau,\varepsilon)+\mu_2(s)H_2(y,s,\tau,\varepsilon)-$$

$$-\frac{\partial}{\partial s}\{\alpha(s)H_0(y,s,\tau,\varepsilon)\}+\frac{1}{2}\frac{\partial^2}{\partial s^2}\left\{\beta^2(s)H_0(y,s,\tau,\varepsilon)\right\},$$

$$-\varepsilon x'(\tau)\frac{\partial H_1(y,s,\tau,\varepsilon)}{\partial y}+\mu_1(s)H_1(y,s,\tau,\varepsilon)=$$

$$=-\varepsilon\lambda\frac{\partial H_1(y,s,\tau,\varepsilon)}{\partial y}+(x+\varepsilon y)H_0(y,s,\tau,\varepsilon)+\varepsilon x\frac{\partial H_0(y,s,\tau,\varepsilon)}{\partial y}-$$

$$-\frac{\partial}{\partial s}\{\alpha(s)H_1(y,s,\tau,\varepsilon)\}+\frac{1}{2}\frac{\partial^2}{\partial s^2}\{\beta^2(s)H_1(y,s,\tau,\varepsilon)\}+o(\varepsilon),$$

$$-\varepsilon x'(\tau)\frac{\partial H_2(y,s,\tau,\varepsilon)}{\partial y}+\mu_2(s)H_2(y,s,\tau,\varepsilon)=$$

$$=\theta H_0(y,s,\tau,\varepsilon)-\varepsilon\lambda\frac{\partial H_2(y,s,\tau,\varepsilon)}{\partial y}-$$

$$-\frac{\partial}{\partial s}\{\alpha(s)H_2(y,s,\tau,\varepsilon)\}+\frac{1}{2}\frac{\partial^2}{\partial s^2}\{\beta^2(s)H_2(y,s,\tau,\varepsilon)\}+o(\varepsilon). \qquad [17.13]$$

By summing the equations of system [17.13] by k, integrating by s and taking the assumption

$$\left(-\alpha(s)\sum_{k=0}^{2}H_k(y,s,\tau,\varepsilon)+\frac{1}{2}\frac{\partial}{\partial s}\left\{\beta^2(s)\sum_{k=0}^{2}H_k(y,s,\tau,\varepsilon)\right\}\right)\Bigg|_{s=-\infty}^{+\infty}=0, \qquad [17.14]$$

into account, we get

$$-\varepsilon x'(\tau)\frac{\partial}{\partial y}\left\{\sum_{k=0}^{2}\int_{-\infty}^{+\infty}H_k(y,s,\tau,\varepsilon)ds\right\}=\varepsilon\frac{\partial}{\partial y}\left\{x\int_{-\infty}^{+\infty}H_0(y,s,\tau,\varepsilon)ds-\right.$$

$$\left.-\lambda\int_{-\infty}^{+\infty}H_2(y,s,\tau,\varepsilon)ds-\lambda\int_{-\infty}^{+\infty}H_1(y,s,\tau,\varepsilon)ds\right\}+o(\varepsilon).$$

We divide both sides of the obtained equation by ε, perform the limit transition and take [17.3] into account, in order to obtain

$$-x'(\tau)\sum_{k=0}^{2}\int_{-\infty}^{+\infty}Q_k(x,s)ds\frac{\partial H(y,\tau)}{\partial y}=\left\{x\int_{-\infty}^{+\infty}Q_0(x,s)ds-\lambda\int_{-\infty}^{+\infty}Q_1(x,s)ds-\right.$$

$$\left.-\lambda\int_{-\infty}^{+\infty}Q_2(x,s)ds\right\}\frac{\partial H(y,\tau)}{\partial y}.$$

Taking [17.5] and [17.6] into account, we obtain

$$\left\{x'(\tau)+xR_0(x)-\lambda(R_1(x)+R_2(x))\right\}\frac{\partial H(y,\tau)}{\partial y}=0\,.$$

The function $x=x(\tau)$ is a solution of an ordinary differential equation:

$$x'(\tau)=-xR_0(x)+\lambda(R_1(x)+R_2(x))=\lambda-(\lambda+x)R_0(x)\,. \qquad [17.15]$$

17.4. Deviation of the number of applications in the system

Let us now consider the process $y(\tau)=\lim\limits_{\varepsilon\to 0}\left(\left(\varepsilon^2 i(\tau/\varepsilon^2)-x(\tau)\right)/\varepsilon\right)$. This process characterizes the deviation of the number of applications in the system. We prove that it is a diffusion autoregression process.

Let us denote the right-hand side of the differential equation [17.15] as $A(x)$:

$$A(x)=\lambda-(\lambda+x)R_0(x) \qquad [17.16]$$

The solution $H_k(y,s,\tau,\varepsilon)$ of system [17.13] can be written in the form

$$H_k(y,s,\tau,\varepsilon)=Q_k(x,s)H(y,\tau)+\varepsilon h_k(y,s,\tau)+o(\varepsilon). \qquad [17.17]$$

We find the kind of functions $h_k(y,s,\tau)$. The system [17.13] can be written in the form

$$-(\theta+\lambda+x)H_0(y,s,\tau,\varepsilon)-\varepsilon yH_0(y,s,\tau,\varepsilon)+\mu_1(s)H_1(y,s,\tau,\varepsilon)+$$

$$+\mu_2(s)H_2(y,s,\tau,\varepsilon)-\frac{\partial}{\partial s}\{\alpha(s)H_0(y,s,\tau,\varepsilon)\}+$$

$$+\frac{1}{2}\frac{\partial^2}{\partial s^2}\left\{\beta^2(s)H_0(y,s,\tau,\varepsilon)\right\}=-\varepsilon x'(\tau)\frac{\partial H_0(y,s,\tau,\varepsilon)}{\partial y},$$

$$-\mu_1(s)H_1(y,s,\tau,\varepsilon)+(\lambda+x)H_0(y,s,\tau,\varepsilon)+\varepsilon yH_0(y,s,\tau,\varepsilon)-$$

$$-\frac{\partial}{\partial s}\{\alpha(s)H_1(y,s,\tau,\varepsilon)\}+\frac{1}{2}\frac{\partial^2}{\partial s^2}\left\{\beta^2(s)H_1(y,s,\tau,\varepsilon)\right\}=$$

$$=-\varepsilon\frac{\partial}{\partial y}\{(x'(\tau)-\lambda)H_1(y,s,\tau,\varepsilon)+xH_0(y,s,\tau,\varepsilon)\}+o(\varepsilon),$$

$$-\mu_2(s)H_2(y,s,\tau,\varepsilon)+\theta H_0(y,s,\tau,\varepsilon)-$$

$$-\frac{\partial}{\partial s}\{\alpha(s)H_2(y,s,\tau,\varepsilon)\}+\frac{1}{2}\frac{\partial^2}{\partial s^2}\{\beta^2(s)H_2(y,s,\tau,\varepsilon)\}=$$

$$=-\varepsilon\frac{\partial}{\partial y}\{(x'(\tau)-\lambda)H_2(y,s,\tau,\varepsilon)\}+o(\varepsilon).$$

By substituting the decomposition [17.17] into this system, taking [17.4] into account and dividing all the equations by ε, the resulting system can be written in the following form:

$$-(\theta+\lambda+x)h_0(y,s,\tau)+\mu_1(s)h_1(y,s,\tau)+\mu_2(s)h_2(y,s,\tau)-$$

$$-\frac{\partial}{\partial s}\{\alpha(s)h_0(y,s,\tau)\}+\frac{1}{2}\frac{\partial^2}{\partial s^2}\{\beta^2(s)h_0(y,s,\tau)\}=$$

$$=Q_0(x,s)yH(y,\tau)-x'(\tau)Q_0(x,s)\frac{\partial H(y,\tau)}{\partial y},$$

$$-\mu_1(s)h_1(y,s,\tau)+(\lambda+x)h_0(y,s,\tau)-$$

$$-\frac{\partial}{\partial s}\{\alpha(s)h_1(y,s,\tau)\}+\frac{1}{2}\frac{\partial^2}{\partial s^2}\{\beta^2(s)h_1(y,s,\tau)\}=$$

$$=-Q_0(x,s)yH(y,\tau)\;-((x'(\tau)-\lambda)Q_1(x,s)+xQ_0(x,s))\frac{\partial H(y,\tau)}{\partial y},$$

$$-\mu_2h_2(y,s,\tau)+\theta h_0(y,s,\tau)-\frac{\partial}{\partial s}\{\alpha(s)h_2(y,s,\tau)\}+$$

$$+\frac{1}{2}\frac{\partial^2}{\partial s^2}\{\beta^2(s)h_2(y,s,\tau)\}=-(x'(\tau)-\lambda)Q_2(x,s)\frac{\partial H(y,\tau)}{\partial y} \qquad [17.18]$$

We will find the solution of system [17.18] in the following form:

$$h_k(y,s,\tau)=h_k^{(1)}(x,s)\frac{\partial H(y,\tau)}{\partial y}+h_k^{(2)}(x,s)yH(y,\tau). \qquad [17.19]$$

We substitute [17.19] into [17.18] and present the system in the form of two systems:

$$-(\theta+\lambda+x)h_0^{(1)}(x,s)+\mu_1(s)h_1^{(1)}(x,s)+\mu_2(s)h_2^{(1)}(x,s)+$$

$$-\frac{\partial}{\partial s}\left\{\alpha(s)h_0^{(1)}(x,s)\right\}+\frac{1}{2}\frac{\partial^2}{\partial s^2}\left\{\beta^2(s)h_0^{(1)}(x,s)\right\}=-x'(\tau)Q_0(x,s),$$

$$-\mu_1(s)h_1^{(1)}(x,s)+(\lambda+x)h_0^{(1)}(x,s)-\frac{\partial}{\partial s}\left\{\alpha(s)h_1^{(1)}(x,s)\right\}+$$

$$+\frac{1}{2}\frac{\partial^2}{\partial s^2}\left\{\beta^2(s)h_1^{(1)}(x,s)\right\}=-(x'(\tau)-\lambda)Q_1(x,s)-xQ_0(x,s),$$

$$-\mu_2(s)h_2^{(1)}(x,s)+\theta h_1^{(1)}(x,s)-\frac{\partial}{\partial s}\{\alpha(s)h_2(x,s)\}+$$

$$+\frac{1}{2}\frac{\partial^2}{\partial s^2}\left\{\beta^2(s)h_2^{(1)}(x,s)\right\}=-(x'(\tau)-\lambda)Q_2(x,s) \qquad [17.20]$$

and

$$-(\theta+\lambda+x)h_0^{(2)}(x,s)+\mu_1(s)h_1^{(2)}(x,s)+\mu_2(s)h_2^{(2)}(x,s)-$$

$$-\frac{\partial}{\partial s}\left\{\alpha(s)h_0^{(2)}(x,s)\right\}+\frac{1}{2}\frac{\partial^2}{\partial s^2}\left\{\beta^2(s)h_0^{(2)}(x,s)\right\}=Q_0(x,s),$$

$$-\mu_1(s)h_1^{(2)}(x,s)+(\lambda+x)h_0^{(2)}(x,s)-\frac{\partial}{\partial s}\left\{\alpha(s)h_1^{(2)}(x,s)\right\}+$$

$$+\frac{1}{2}\frac{\partial^2}{\partial s^2}\left\{\beta^2(s)h_1^{(2)}(x,s)\right\}=-Q_0(x,s),$$

$$-\mu_2(s)h_2^{(2)}(x,s)+\theta h_0^{(2)}(x,s)-$$

$$-\frac{\partial}{\partial s}\{\alpha(s)h_2^{(2)}(x,s)\}+\frac{1}{2}\frac{\partial^2}{\partial s^2}\{\beta^2(s)h_2^{(2)}(x,s)\}=0 \qquad [17.21]$$

Differentiate the system [17.4] by x. Consequently, the solution $h_k^{(2)}(x,s)$ of system [17.21] has the form

$$h_k^{(2)}(x,s)=\frac{\partial Q_k(x,s)}{\partial x} \qquad [17.22]$$

By taking into account [17.22] and [17.19], decomposition [17.17] has the form

$$H_k(y,s,\tau,\varepsilon)=Q_k(x,s)H(y,\tau)+\varepsilon h_k^{(1)}(x,s)\frac{\partial H(y,\tau)}{\partial y}+$$

$$+\varepsilon yH(y,\tau)\frac{\partial Q_k(x,s)}{\partial x}+o(\varepsilon) \qquad [17.23]$$

We now find the type of function $H(y,\tau)$. The functions on the right-hand side of system [17.1] are expanded in a series in increments of the argument y up to $o(\varepsilon^2)$, in order to obtain

$$\varepsilon^2\frac{\partial H_0(y,s,\tau,\varepsilon)}{\partial\tau}-\varepsilon x'(\tau)\frac{\partial H_0(y,s,\tau,\varepsilon)}{\partial y}+(\theta+\lambda+x+\varepsilon y)H_0(y,s,\tau,\varepsilon)=$$

$$=\mu_1(s)H_1(y,s,\tau,\varepsilon)+\mu_2(s)H_2(y,s,\tau,\varepsilon)-$$

$$-\frac{\partial}{\partial s}\{\alpha(s)H_0(y,s,\tau,\varepsilon)\}+\frac{1}{2}\frac{\partial^2}{\partial s^2}\{\beta^2(s)H_0(y,s,\tau,\varepsilon)\},$$

$$\varepsilon^2\frac{\partial H_1(y,s,\tau,\varepsilon)}{\partial\tau}-\varepsilon x'(\tau)\frac{\partial H_1(y,s,\tau,\varepsilon)}{\partial y}+(\lambda+\mu_1(s))H_1(y,s,\tau,\varepsilon)=$$

$$=\lambda H_1(y,s,\tau,\varepsilon)-\varepsilon\lambda\frac{\partial H_1(y,s,\tau,\varepsilon)}{\partial y}+\lambda\frac{\varepsilon^2}{2}\frac{\partial^2 H_1(y,s,\tau,\varepsilon)}{\partial y^2}+$$

$$+(\lambda+x+\varepsilon y)H_0(y,s,\tau,\varepsilon)+\varepsilon\frac{\partial}{\partial y}\{(x+\varepsilon y)H_0(y,s,\tau,\varepsilon)\}+$$

$$+x\frac{\varepsilon^2}{2}\frac{\partial^2 H_0(y,s,\tau,\varepsilon)}{\partial y^2}-\frac{\partial}{\partial s}\{\alpha(s)H_1(y,s,\tau,\varepsilon)\}+$$

$$+\frac{1}{2}\frac{\partial^2}{\partial s^2}\left\{\beta^2(s)H_1(y,s,\tau,\varepsilon)\right\}+o(\varepsilon^2),$$

$$\varepsilon^2\frac{\partial H_2(y,s,\tau,\varepsilon)}{\partial \tau}-\varepsilon x'(\tau)\frac{\partial H_2(y,s,\tau,\varepsilon)}{\partial y}+(\lambda+\mu_2(s))H_2(y,s,\tau,\varepsilon)=$$

$$=\theta H_0(y,s,\tau,\varepsilon)+\lambda H_2(y,s,\tau,\varepsilon)-\lambda\varepsilon\frac{\partial H_2(y,s,\tau,\varepsilon)}{\partial y}+$$

$$+\lambda\frac{\varepsilon^2}{2}\frac{\partial^2 H_2(y,s,\tau,\varepsilon)}{\partial y^2}-\frac{\partial}{\partial s}\{\alpha(s)H_2(y,s,\tau,\varepsilon)\}+$$

$$+\frac{1}{2}\frac{\partial^2}{\partial s^2}\left\{\beta^2(s)H_2(y,s,\tau,\varepsilon)\right\}+o(\varepsilon^2) \qquad [17.24]$$

By summing the equations of system [17.24] by k, substituting the expansion of functions $H_k(y,s,\tau,\varepsilon)$ into the resulting system in the form [17.23] and taking the notation [17.6] into account, we obtain

$$\varepsilon^2 r(s)\frac{\partial H(y,\tau)}{\partial \tau}-\varepsilon x'(\tau)r(s)\frac{\partial H(y,\tau)}{\partial y}-$$

$$-\varepsilon^2 x'(\tau)\frac{\partial}{\partial x}\left\{\sum_{k=0}^{2}Q_k(x,s)\right\}\frac{\partial\{yH(y,\tau)\}}{\partial y}-\varepsilon^2 x'(\tau)\sum_{k=0}^{2}h_k^{(1)}(x,s)\frac{\partial^2 H(y,\tau)}{\partial y^2}=$$

$$=-\varepsilon\left(-xQ_0(x,s)+\lambda Q_1(x,s)+\lambda Q_2(x,s)\right)\frac{\partial H(y,\tau)}{\partial y}-$$

$$-\varepsilon^2\left(-Q_0(x,s)-x\frac{\partial Q_0(x,s)}{\partial x}+\lambda\frac{\partial Q_1(x,s)}{\partial x}+\lambda\frac{\partial Q_2(x,s)}{\partial x}\right)\frac{\partial\{yH(y,\tau)\}}{\partial y}+$$

$$+\frac{\varepsilon^2}{2}[xQ_0(x,s)+\lambda Q_1(x,s)+\lambda Q_2(x,s)+2\left(xh_0^{(1)}(x,s)-\lambda h_1^{(1)}(x,s)-\right.$$

$$-\lambda h_2^{(1)}(x,s)\big)\Big]\frac{\partial^2 H(y,\tau)}{\partial y^2}-\frac{\partial}{\partial s}\left\{\alpha(s)\sum_{k=0}^{2}H_k(y,s,\tau,\varepsilon)\right\}+$$

$$+\frac{1}{2}\frac{\partial^2}{\partial s^2}\left\{\beta^2(s)\sum_{k=0}^{2}H_k(y,s,\tau,\varepsilon)\right\}+o(\varepsilon^2) \qquad [17.25]$$

We integrate the left and right sides of the equation [17.25] by s, use the condition [17.7], the designation [17.6] and also denote

$$\int_{-\infty}^{+\infty} h_k^{(1)}(x,s)ds = h_k^{(1)}(x),\ \sum_{k=0}^{2} h_k^{(1)}(x) = h^{(1)}(x) \qquad [17.26]$$

By taking [17.14] and [17.15] into account, as well as dividing both sides of the equation by ε^2, we obtain

$$\frac{\partial H(y,\tau)}{\partial \tau}=-\left(-R_0(x)-x\frac{\partial R_0(x)}{\partial x}+\lambda\frac{\partial}{\partial x}\{R_1(x)+R_2(x)\}\right)\frac{\partial\{yH(y,\tau)\}}{\partial y}+$$

$$+\frac{1}{2}\big[xR_0(x)+\lambda R_1(x)+\lambda R_2(x)+$$

$$+2\left(xh_0^{(1)}(x)-(\lambda+x)R_0(x)h^{(1)}(x)\right)\Big]\frac{\partial^2 H(y,\tau)}{\partial y^2} \qquad [17.27]$$

We derived Fokker's and Planck's equation for the probability density $H(y,\tau)$. The drift coefficient of equation [17.27] is a derivative of the right-hand side of the differential equation [17.15]:

$$A_x'(x)=-R_0(x)-x\frac{\partial R_0(x)}{\partial x}+\lambda\frac{\partial}{\partial x}\{R_1(x)+R_2(x)\}=$$

$$=\frac{\partial}{\partial x}\{-xR_0(x)+\lambda(R_1(x)+R_2(x))\}=\frac{\partial}{\partial x}\{\lambda-(\lambda+x)R_0(x)\} \qquad [17.28]$$

By taking [17.26] into account, we denote the diffusion coefficient as follows:

$$B^2(x)=\lambda-(\lambda-x)R_0(x)+2xh_0^{(1)}(x)-2(\lambda+x)R_0(x)h^{(1)}(x) \qquad [17.29]$$

The function $H(y,\tau)$ is a probability density of values certain diffusion process $y(\tau)$. The process $y(\tau)$ satisfies the stochastic differential equation:

$$dy(\tau) = A'_x(x)y(\tau)d\tau + B(x)dw(\tau) \qquad [17.30]$$

where $w(\tau)$ is a standard Wiener process; therefore, the process $y(\tau)$ is an autoregression process.

17.5. Probability distribution density of device states

The random process $z(\tau) = x(\tau) + \varepsilon y$ that approximates the process of changing the number of customers in the system $\varepsilon^2 i(\tau/\varepsilon^2)$ is a homogeneous diffusion process.

We differentiate $z(\tau)$ by τ:

$$dz(\tau) = x'(\tau)d\tau + \varepsilon dy. \qquad [17.31]$$

Taking [17.15] and [17.30] into account

$$dz(\tau) = [-xR_0(x) + \lambda(R_1(x) + R_2(x))]d\tau +$$

$$+\varepsilon y \frac{\partial}{\partial x}\{-xR_0(x) + \lambda(R_1(x) + R_2(x))\}d\tau + \varepsilon B(x)dw(\tau).$$

The right-hand side of the equation has a series expansion in increments εy of the argument x. We can then write

$$dz(\tau) = [-(x+\varepsilon y)R_0(x+\varepsilon y) + \lambda(R_1(x+\varepsilon y) + R_2(x+\varepsilon y))]d\tau +$$

$$+\varepsilon B(z-\varepsilon y)dw(\tau).$$

With the accuracy $o(\varepsilon)$, we have

$$dz(\tau) = [-zR_0(z) + \lambda(R_1(z) + R_2(z))]d\tau + \varepsilon B(z)dw(\tau) + o(\varepsilon).$$

By taking [17.16] into account, we obtain

$$dz(\tau) = A(z)d\tau + \varepsilon B(z)dw(\tau) + o(\varepsilon).$$

Thus, $z(\tau)$ is a homogeneous diffusion process with a drift coefficient $A(z)$ and a diffusion coefficient $\varepsilon^2 B^2(z)$, which is determined with an accuracy $o(\varepsilon)$ to a stochastic differential equation of the form [17.31].

Let us denote $F(z,\tau)$ as the probability density of the process values $z(\tau)$. We can write Fokker's and Planck's equation for the density of this process in the form:

$$\frac{\partial F(z,\tau)}{\partial \tau} = -\frac{\partial}{\partial z}\{A(z)F(z,\tau)\} + \frac{\varepsilon^2}{2}\frac{\partial^2}{\partial z^2}\{B^2(z)F(z,\tau)\}.$$

Consider the functioning of the process $z(\tau)$ in a stationary mode $F(z,\tau) \equiv F(z)$. The stable distribution can be found from the equation

$$0 = -\frac{\partial}{\partial z}\{A(z)F(z)\} + \frac{\varepsilon^2}{2}\frac{\partial^2}{\partial z^2}\{B^2(z)F(z)\}.$$

This equation is a homogeneous differential equation that has the solution

$$F(z) = \frac{1}{B^2(z)} e^{\frac{2}{\varepsilon^2}\int_0^z \frac{A(u)}{B^2(u)}du} \Bigg/ \int_0^\infty \frac{1}{B^2(z)} e^{\frac{2}{\varepsilon^2}\int_0^z \frac{A(u)}{B^2(u)}du} dz. \quad [17.32]$$

17.6. Conclusion

In this chapter, for the presented model of the retrial queuing system in a diffusion environment, we found the asymptotic average of the normalized number of calls in the system in the form [17.15], the probability distribution of the device states [17.12] and the deviation from the average, which are determined by the stochastic equation [17.30]. The approximation of the number of applications in the system was carried out by a homogeneous diffusion process. The probability density of the process values was found in the form [17.32]. The results can be used in servicing systems, such as call centers, in order to increase efficiency.

17.7. References

Artalejo, J.R. and Gomez-Corral, A. (2008). *Retrial Queueing Systems: A Computational Approach*. Springer, Heidelberg.

Nazarov, A., Phung-Duc, T., Paul, S. (2019). Retrial asymptotics for a single server queue with two-way communication and Markov Modulated Poison Input. *Journal of Systems Science and Systems Engineering*, 28(2), 181–193.

Neuts, M.P. (1971). A queue subject to extraneous phase changes. *Advances in Applied Probability*, 3, 78–119.

Neuts, M.P. (1978). Further results of the M/M/1 queue with randomly varying rates. *Opsearch*, 15, 139–157.

Purdue, P. (1974). The M/M/1 queue in a Markovian environment. *Operations Research*, 22, 562–569.

Yechiali, U. (1973). A queuing-type birth-and-death process defined on a continuous-time Markov Chain. *Operations Research*, 21, 604–609.

Yechiali, U. and Naor P. (1971). Queuing problems with heterogeneous arrivals and services. *Operations Research*, 19, 722–734.

List of Authors

Mohammed ALBUHAYRI
Division of Applied Mathematics
Mälardalen University
Västerås
Sweden

Roberto ASCARI
Department of Economics,
Management and Statistics (DEMS)
University of Milano-Bicocca
Milan
Italy

Mauro BARONE
Risk Analysis and Tax Compliance
Research Unit
Italian Revenue Agency
Rome
Italy

Ekaterina BULINSKAYA
Faculty of Mechanics and
Mathematics
Lomonosov Moscow State University
Russia

Mark Anthony CARUANA
Department of Statistics and
Operations Research
University of Malta
Msida
Malta

Pierre DEVOLDER
Louvain Institute of Data Analysis and
Modeling in Economics and Statistics
Catholic University of Louvain
Belgium

Keivan DIAKITE
Louvain Institute of Data Analysis and
Modeling in Economics and Statistics
Catholic University of Louvain
Belgium

Agnese M. DI BRISCO
Department of Economics,
Management and Statistics (DEMS)
University of Milano-Bicocca
Milan
Italy

Yannis DIMOTIKALIS
Department of Management Science and Technology
Hellenic Mediterranean University
Heraklion
Crete
Greece

Christopher ENGSTRÖM
Division of Applied Mathematics
Mälardalen University
Västerås
Sweden

Francesco GIORDANO
Department of Economics and Statistics (DiSES)
University of Salerno
Fisciano
Italy

Azam A. IMOMOV
Department of Mathematics
Karshi State University
Karshi City
Uzbekistan

Petr JIZBA
Department of Physics
Czech Technical University in Prague
Czech Republic

Godwin KAKUBA
Department of Mathematics
Makerere University
Kampala
Uganda

Alex KARAGRIGORIOU
Department of Statistics and Actuarial-Financial Mathematics
University of the Aegean
Samos
Greece

Marianna KOUKLI
Department of History and Ethnology
Democritus University of Thrace
Komotini
Greece

Karl LUNDENGÅRD
Division of Applied Mathematics
Mälardalen University
Västerås
Sweden

Anatoliy MALYARENKO
Division of Applied Mathematics
Mälardalen University
Västerås
Sweden

John Magero MANGO
Department of Mathematics
Makerere University
Kampala
Uganda

Aggeliki MARAGKAKI
Health Care Management
Hellenic Open University
Heraklion
Crete
Greece

George MATALLIOTAKIS
Health Care Management
Hellenic Open University
Heraklion
Crete
Greece

Sonia MIGLIORATI
Department of Economics,
Management and Statistics (DEMS)
University of Milano-Bicocca
Milan
Italy

Asaph Keikara MUHUMUZA
Division of Applied Mathematics
Mälardalen University
Västerås
Sweden

Jean-Paul MURARA
Division of Applied Mathematics
Mälardalen University
Västerås
Sweden

Ying NI
Division of Applied Mathematics
Mälardalen University
Västerås
Sweden

Marcella NIGLIO
Department of Economics and
Statistics (DiSES)
University of Salerno
Fisciano
Italy

Hossein NOHROUZIAN
Division of Applied Mathematics
Mälardalen University
Västerås
Sweden

Andrea ONGARO
Department of Economics,
Management and Statistics (DEMS)
University of Milano-Bicocca
Milan
Italy

Abderrahim OULIDI
Department of Statistics and Actuarial
International University of Rabat
Morroco

Christina PARPOULA
Department of Psychology
Panteion University of Social and
Political Sciences
Athens
Greece

Stefano PISANI
Risk Analysis and Tax Compliance
Research Unit
Italian Revenue Agency
Rome
Italy

Martin PROKŠ
Department of Physics
Czech Technical University
in Prague
Czech Republic

Milica RANCIC
Division of Applied Mathematics
Mälardalen University
Västerås
Sweden

Sergei SILVESTROV
Division of Applied Mathematics
Mälardalen University
Västerås
Sweden

Charilaos SKIADAS
Department of Mathematics and
Computer Science
Hanover College
Indiana
USA

Christos H. SKIADAS
ManLab
Department of Production
Engineering and Management
Technical University of Crete
Chania
Greece

Andrea SPINGOLA
Risk Analysis and Tax Compliance
Research Unit
Italian Revenue Agency
Rome
Italy

Finnan TEWOLDE
Division of Applied Mathematics
Mälardalen University
Västerås
Sweden

Erkin E. TUKHTAEV
Department of Mathematics
Karshi State University
Karshi City
Uzbekistan

Viacheslav VAVILOV
Department of Software Engineering
National Research Tomsk
State University
Russia

Konstantinos ZAFEIRIS
Department of History and Ethnology
Democritus University of Thrace
Komotini
Greece

Jiahui ZHANG
Nasdaq AB
Stockholm
Sweden

Index

N, O, P

Q, R

S

T

V, W

Other titles from

iSTE

in

Innovation, Entrepreneurship and Management

2021

BOBILLIER CHAUMON Marc-Eric
Digital Transformations in the Challenge of Activity and Work: Understanding and Supporting Technological Changes
(Technological Changes and Human Resources Set – Volume 3)

2020

ACH Yves-Alain, RMADI-SAÏD Sandra
Financial Information and Brand Value: Reflections, Challenges and Limitations

ANDREOSSO-O'CALLAGHAN Bernadette, DZEVER Sam, JAUSSAUD Jacques, TAYLOR Robert
Sustainable Development and Energy Transition in Europe and Asia
(Innovation and Technology Set – Volume 9)

BEN SLIMANE Sonia, M'HENNI Hatem
Entrepreneurship and Development: Realities and Future Prospects
(Smart Innovation Set – Volume 30)

CHOUTEAU Marianne, FOREST Joëlle, NGUYEN Céline
Innovation for Society: The P.S.I. Approach
(Smart Innovation Set – Volume 28)

CORON Clotilde
Quantifying Human Resources: Uses and Analysis
(Technological Changes and Human Resources Set – Volume 2)

CORON Clotilde, GILBERT Patrick
Technological Change
(Technological Changes and Human Resources Set – Volume 1)

CERDIN Jean-Luc, PERETTI Jean-Marie
The Success of Apprenticeships: Views of Stakeholders on Training and Learning
(Human Resources Management Set – Volume 3)

DELCHET-COCHET Karen
Circular Economy: From Waste Reduction to Value Creation
(Economic Growth Set – Volume 2)

DIDAY Edwin, GUAN Rong, SAPORTA Gilbert, WANG Huiwen
Advances in Data Science
(Big Data, Artificial Intelligence and Data Analysis Set – Volume 4)

DOS SANTOS PAULINO Victor
Innovation Trends in the Space Industry
(Smart Innovation Set – Volume 25)

GASMI Nacer
Corporate Innovation Strategies: Corporate Social Responsibility and Shared Value Creation
(Smart Innovation Set – Volume 33)

GOGLIN Christian
Emotions and Values in Equity Crowdfunding Investment Choices 1: Transdisciplinary Theoretical Approach

GUILHON Bernard
Venture Capital and the Financing of Innovation
(Innovation Between Risk and Reward Set – Volume 6)

LATOUCHE Pascal
Open Innovation: Human Set-up
(Innovation and Technology Set – Volume 10)

LIMA Marcos
Entrepreneurship and Innovation Education: Frameworks and Tools
(Smart Innovation Set – Volume 32)

MACHADO Carolina, DAVIM J. Paulo
Sustainable Management for Managers and Engineers

MAKRIDES Andreas, KARAGRIGORIOU Alex, SKIADAS Christos H.
Data Analysis and Applications 3: Computational, Classification, Financial, Statistical and Stochastic Methods
(Big Data, Artificial Intelligence and Data Analysis Set – Volume 5)
Data Analysis and Applications 4: Financial Data Analysis and Methods
(Big Data, Artificial Intelligence and Data Analysis Set – Volume 6)

MASSOTTE Pierre, CORSI Patrick
Complex Decision-Making in Economy and Finance

MEUNIER François-Xavier
Dual Innovation Systems: Concepts, Tools and Methods
(Smart Innovation Set – Volume 31)

MICHAUD Thomas
Science Fiction and Innovation Design (Innovation in Engineering and Technology Set – Volume 6)

MONINO Jean-Louis
Data Control: Major Challenge for the Digital Society
(Smart Innovation Set – Volume 29)

MORLAT Clément
Sustainable Productive System: Eco-development versus Sustainable Development
(Smart Innovation Set – Volume 26)

SAULAIS Pierre, ERMINE Jean-Louis
Knowledge Management in Innovative Companies 2: Understanding and Deploying a KM Plan within a Learning Organization
(Smart Innovation Set – Volume 27)

2019

AMENDOLA Mario, GAFFARD Jean-Luc
Disorder and Public Concern Around Globalization

BARBAROUX Pierre
Disruptive Technology and Defence Innovation Ecosystems
(Innovation in Engineering and Technology Set – Volume 5)

DOU Henri, JUILLET Alain, CLERC Philippe
Strategic Intelligence for the Future 1: A New Strategic and Operational Approach
Strategic Intelligence for the Future 2: A New Information Function Approach

FRIKHA Azza
Measurement in Marketing: Operationalization of Latent Constructs

FRIMOUSSE Soufyane
Innovation and Agility in the Digital Age
(Human Resources Management Set – Volume 2)

GAY Claudine, SZOSTAK Bérangère L.
Innovation and Creativity in SMEs: Challenges, Evolutions and Prospects
(Smart Innovation Set – Volume 21)

GORIA Stéphane, HUMBERT Pierre, ROUSSEL Benoît
Information, Knowledge and Agile Creativity
(Smart Innovation Set – Volume 22)

HELLER David
Investment Decision-making Using Optional Models
(Economic Growth Set – Volume 2)

HELLER David, DE CHADIRAC Sylvain, HALAOUI Lana, JOUVET Camille
The Emergence of Start-ups
(Economic Growth Set – Volume 1)

HÉRAUD Jean-Alain, KERR Fiona, BURGER-HELMCHEN Thierry
Creative Management of Complex Systems
(Smart Innovation Set – Volume 19)

LATOUCHE Pascal
Open Innovation: Corporate Incubator
(Innovation and Technology Set – Volume 7)

LEHMANN Paul-Jacques
The Future of the Euro Currency

LEIGNEL Jean-Louis, MÉNAGER Emmanuel, YABLONSKY Serge
Sustainable Enterprise Performance: A Comprehensive Evaluation Method

LIÈVRE Pascal, AUBRY Monique, GAREL Gilles
Management of Extreme Situations: From Polar Expeditions to Exploration-Oriented Organizations

MILLOT Michel
Embarrassment of Product Choices 2: Towards a Society of Well-being

N'GOALA Gilles, PEZ-PÉRARD Virginie, PRIM-ALLAZ Isabelle
Augmented Customer Strategy: CRM in the Digital Age

NIKOLOVA Blagovesta
The RRI Challenge: Responsibilization in a State of Tension with Market Regulation
(Innovation and Responsibility Set – Volume 3)

PELLEGRIN-BOUCHER Estelle, ROY Pierre
Innovation in the Cultural and Creative Industries
(Innovation and Technology Set – Volume 8)

PRIOLON Joël
Financial Markets for Commodities

QUINIOU Matthieu
Blockchain: The Advent of Disintermediation

RAVIX Joël-Thomas, DESCHAMPS Marc
Innovation and Industrial Policies
(Innovation between Risk and Reward Set – Volume 5)

ROGER Alain, VINOT Didier
Skills Management: New Applications, New Questions
(Human Resources Management Set – Volume 1)

SAULAIS Pierre, ERMINE Jean-Louis
Knowledge Management in Innovative Companies 1: Understanding and Deploying a KM Plan within a Learning Organization
(Smart Innovation Set – Volume 23)

SERVAJEAN-HILST Romaric
Co-innovation Dynamics: The Management of Client-Supplier Interactions for Open Innovation
(Smart Innovation Set – Volume 20)

SKIADAS Christos H., BOZEMAN James R.
Data Analysis and Applications 1: Clustering and Regression, Modeling-estimating, Forecasting and Data Mining
(Big Data, Artificial Intelligence and Data Analysis Set – Volume 2)
Data Analysis and Applications 2: Utilization of Results in Europe and Other Topics
(Big Data, Artificial Intelligence and Data Analysis Set – Volume 3)

UZUNIDIS Dimitri
Systemic Innovation: Entrepreneurial Strategies and Market Dynamics

VIGEZZI Michel
World Industrialization: Shared Inventions, Competitive Innovations and Social Dynamics
(Smart Innovation Set – Volume 24)

2018

BURKHARDT Kirsten
Private Equity Firms: Their Role in the Formation of Strategic Alliances

CALLENS Stéphane
Creative Globalization
(Smart Innovation Set – Volume 16)

CASADELLA Vanessa
Innovation Systems in Emerging Economies: MINT – Mexico, Indonesia, Nigeria, Turkey
(Smart Innovation Set – Volume 18)

CHOUTEAU Marianne, FOREST Joëlle, NGUYEN Céline
Science, Technology and Innovation Culture
(Innovation in Engineering and Technology Set – Volume 3)

CORLOSQUET-HABART Marine, JANSSEN Jacques
Big Data for Insurance Companies
(Big Data, Artificial Intelligence and Data Analysis Set – Volume 1)

CROS Françoise
Innovation and Society
(Smart Innovation Set – Volume 15)

DEBREF Romain
Environmental Innovation and Ecodesign: Certainties and Controversies
(Smart Innovation Set – Volume 17)

DOMINGUEZ Noémie
SME Internationalization Strategies: Innovation to Conquer New Markets

ERMINE Jean-Louis
Knowledge Management: The Creative Loop
(Innovation and Technology Set – Volume 5)

GILBERT Patrick, BOBADILLA Natalia, GASTALDI Lise,
LE BOULAIRE Martine, LELEBINA Olga
Innovation, Research and Development Management

IBRAHIMI Mohammed
Mergers & Acquisitions: Theory, Strategy, Finance

LEMAÎTRE Denis
Training Engineers for Innovation

LÉVY Aldo, BEN BOUHENI Faten, AMMI Chantal
Financial Management: USGAAP and IFRS Standards
(Innovation and Technology Set – Volume 6)

MILLOT Michel
Embarrassment of Product Choices 1: How to Consume Differently

PANSERA Mario, OWEN Richard
Innovation and Development: The Politics at the Bottom of the Pyramid
(Innovation and Responsibility Set – Volume 2)

RICHEZ Yves
Corporate Talent Detection and Development

SACHETTI Philippe, ZUPPINGER Thibaud
New Technologies and Branding
(Innovation and Technology Set – Volume 4)

SAMIER Henri
Intuition, Creativity, Innovation

TEMPLE Ludovic, COMPAORÉ SAWADOGO Eveline M.F.W.
Innovation Processes in Agro-Ecological Transitions in Developing Countries
(Innovation in Engineering and Technology Set – Volume 2)

UZUNIDIS Dimitri
Collective Innovation Processes: Principles and Practices
(Innovation in Engineering and Technology Set – Volume 4)

VAN HOOREBEKE Delphine
The Management of Living Beings or Emo-management

2017

AÏT-EL-HADJ Smaïl
The Ongoing Technological System
(Smart Innovation Set – Volume 11)

BAUDRY Marc, DUMONT Béatrice
Patents: Prompting or Restricting Innovation?
(Smart Innovation Set – Volume 12)

BÉRARD Céline, TEYSSIER Christine
Risk Management: Lever for SME Development and Stakeholder Value Creation

CHALENÇON Ludivine
Location Strategies and Value Creation of International Mergers and Acquisitions

CHAUVEL Danièle, BORZILLO Stefano
The Innovative Company: An Ill-defined Object
(Innovation between Risk and Reward Set – Volume 1)

CORSI Patrick
Going Past Limits To Growth

D'ANDRIA Aude, GABARRET Inés
Building 21st Century Entrepreneurship
(Innovation and Technology Set – Volume 2)

DAIDJ Nabyla
Cooperation, Coopetition and Innovation
(Innovation and Technology Set – Volume 3)

FERNEZ-WALCH Sandrine
The Multiple Facets of Innovation Project Management
(Innovation between Risk and Reward Set – Volume 4)

FOREST Joëlle
Creative Rationality and Innovation
(Smart Innovation Set – Volume 14)

GUILHON Bernard
Innovation and Production Ecosystems
(Innovation between Risk and Reward Set – Volume 2)

HAMMOUDI Abdelhakim, DAIDJ Nabyla
Game Theory Approach to Managerial Strategies and Value Creation
(Diverse and Global Perspectives on Value Creation Set – Volume 3)

LALLEMENT Rémi
Intellectual Property and Innovation Protection: New Practices and New Policy Issues
(Innovation between Risk and Reward Set – Volume 3)

LAPERCHE Blandine
Enterprise Knowledge Capital
(Smart Innovation Set – Volume 13)

LEBERT Didier, EL YOUNSI Hafida
International Specialization Dynamics
(Smart Innovation Set – Volume 9)

MAESSCHALCK Marc
Reflexive Governance for Research and Innovative Knowledge
(Responsible Research and Innovation Set – Volume 6)

MASSOTTE Pierre
Ethics in Social Networking and Business 1: Theory, Practice and Current Recommendations
Ethics in Social Networking and Business 2: The Future and Changing Paradigms

MASSOTTE Pierre, CORSI Patrick
Smart Decisions in Complex Systems

MEDINA Mercedes, HERRERO Mónica, URGELLÉS Alicia
Current and Emerging Issues in the Audiovisual Industry
(Diverse and Global Perspectives on Value Creation Set – Volume 1)

MICHAUD Thomas
Innovation, Between Science and Science Fiction
(Smart Innovation Set – Volume 10)

PELLÉ Sophie
Business, Innovation and Responsibility
(Responsible Research and Innovation Set – Volume 7)

SAVIGNAC Emmanuelle
The Gamification of Work: The Use of Games in the Workplace

SUGAHARA Satoshi, DAIDJ Nabyla, USHIO Sumitaka
Value Creation in Management Accounting and Strategic Management: An Integrated Approach
(Diverse and Global Perspectives on Value Creation Set –Volume 2)

UZUNIDIS Dimitri, SAULAIS Pierre
Innovation Engines: Entrepreneurs and Enterprises in a Turbulent World
(Innovation in Engineering and Technology Set – Volume 1)

2016

BARBAROUX Pierre, ATTOUR Amel, SCHENK Eric
Knowledge Management and Innovation
(Smart Innovation Set – Volume 6)

BEN BOUHENI Faten, AMMI Chantal, LEVY Aldo
Banking Governance, Performance And Risk-Taking: Conventional Banks Vs Islamic Banks

BOUTILLIER Sophie, CARRÉ Denis, LEVRATTO Nadine
Entrepreneurial Ecosystems (Smart Innovation Set – Volume 2)

BOUTILLIER Sophie, UZUNIDIS Dimitri
The Entrepreneur (Smart Innovation Set – Volume 8)

BOUVARD Patricia, SUZANNE Hervé
Collective Intelligence Development in Business

GALLAUD Delphine, LAPERCHE Blandine
Circular Economy, Industrial Ecology and Short Supply Chains
(Smart Innovation Set – Volume 4)

GUERRIER Claudine
Security and Privacy in the Digital Era
(Innovation and Technology Set – Volume 1)

MEGHOUAR Hicham
Corporate Takeover Targets

MONINO Jean-Louis, SEDKAOUI Soraya
Big Data, Open Data and Data Development
(Smart Innovation Set – Volume 3)

MOREL Laure, LE ROUX Serge
Fab Labs: Innovative User
(Smart Innovation Set – Volume 5)

PICARD Fabienne, TANGUY Corinne
Innovations and Techno-ecological Transition
(Smart Innovation Set – Volume 7)

2015

CASADELLA Vanessa, LIU Zeting, DIMITRI Uzunidis
Innovation Capabilities and Economic Development in Open Economies
(Smart Innovation Set – Volume 1)

CORSI Patrick, MORIN Dominique
Sequencing Apple's DNA

CORSI Patrick, NEAU Erwan
Innovation Capability Maturity Model

FAIVRE-TAVIGNOT Bénédicte
Social Business and Base of the Pyramid

GODÉ Cécile
Team Coordination in Extreme Environments

MAILLARD Pierre
Competitive Quality and Innovation

MASSOTTE Pierre, CORSI Patrick
Operationalizing Sustainability

MASSOTTE Pierre, CORSI Patrick
Sustainability Calling

2014

DUBÉ Jean, LEGROS Diègo
Spatial Econometrics Using Microdata

LESCA Humbert, LESCA Nicolas
Strategic Decisions and Weak Signals

2013

HABART-CORLOSQUET Marine, JANSSEN Jacques, MANCA Raimondo
VaR Methodology for Non-Gaussian Finance

2012

DAL PONT Jean-Pierre
Process Engineering and Industrial Management

MAILLARD Pierre
Competitive Quality Strategies

POMEROL Jean-Charles
Decision-Making and Action

SZYLAR Christian
UCITS Handbook

2011

LESCA Nicolas
Environmental Scanning and Sustainable Development

LESCA Nicolas, LESCA Humbert
Weak Signals for Strategic Intelligence: Anticipation Tool for Managers

MERCIER-LAURENT Eunika
Innovation Ecosystems

2010

SZYLAR Christian
Risk Management under UCITS III/IV

2009

COHEN Corine
Business Intelligence

ZANINETTI Jean-Marc
Sustainable Development in the USA

2008

CORSI Patrick, DULIEU Mike
The Marketing of Technology Intensive Products and Services

DZEVER Sam, JAUSSAUD Jacques, ANDREOSSO Bernadette
Evolving Corporate Structures and Cultures in Asia: Impact of Globalization

2007

AMMI Chantal
Global Consumer Behavior

2006

BOUGHZALA Imed, ERMINE Jean-Louis
Trends in Enterprise Knowledge Management

CORSI Patrick *et al.*
Innovation Engineering: the Power of Intangible Networks